Fire Safety and Loss Prevention

AMERICAN SOCIETY FOR INDUSTRIAL SECURITY
1625 PRINCE STREET
ALEXANDRIA, VA 22314
(703) 519–6200

Fire Safety and Loss Prevention

Kevin A. Cassidy

Butterworth-Heinemann

Boston   London   Oxford   Singapore   Sydney   Toronto   Wellington

Recognizing the importance of preserving what has been
written, it is the policy of Butterworth-Heinemann to have
the books it publishes printed on acid-free paper, and we
exert our best efforts to that end.

Library of Congress Cataloging-in-Publication Data

Cassidy, Kevin A., 1953–
    Fire Safety and loss prevention / Kevin A. Cassidy.
        p.   cm.
    Includes bibliographical references and index.
    ISBN 0-7506-9039-9
    1. Fire prevention.   2.   Burglary
protection.   I.   Title.
TH9145.C38   1992
363.37′7—dc20   91-28930
                                                        CIP

British Library Cataloguing in Publication Data

Cassidy, Kevin A.
    Fire safety and loss prevention.
    I.   Title
    628.92

ISBN 0-7506-9039-9

Butterworth-Heinemann
80 Montvale Avenue
Stoneham, MA 02180

10 9 8 7 6 5 4 3 2 1

Printed in the United States of America

My special thanks to Debra, Kevin II, Brett, and Lauren

In memory of John Kenneth Law

Contents

Preface

Fire safety, security, and loss prevention must interface continually in order for states, cities, businesses, and the economy to interact successfully. Nowadays, much emphasis is placed on educating the public regarding fire safety and survival during a fire. This book explores the ramifications of fire prevention and a security department's responsibility prior to, during, and after a fire. At first glance, the responsibilities seem simple enough. However, these duties become more complex when one considers the various building codes, life safety codes, and fire codes associated with a building's design.

Security directors, managers, and supervisors, as well as building, hospital, retail establishment, and commercial building owners will gain vital information from this book. Conflicts constantly arise between the importance of fire safety and security. Which is more crucial to your establishment? Often, security priorities must be lowered in deference to fire safety mandates. The argument that plagues management is—What takes top priority? Fire safety or security? The answer is fire safety. The majority of fire safety regulations are mandated; compliance is a matter of obeying the law, whereas most security regulations are not required by law.

It is crucial for the reader to remember that this is not a technical book. The ideas and suggestions are presented for you to analyze and maybe incorporate in your facility. Many times, a colleague has asked me or I have asked a colleague, where do I obtain information regarding specific codes? You can spend many agonizing hours searching for information. In this book, I list the fifty states plus the District of Columbia, Puerto Rico, and the Virgin Islands, and what is required regarding building, fire, and life safety codes, as well as the enforcement of these codes.

Throughout the text, you will come across material pertaining to *codes*. Check to see which codes apply to your state or city. New York codes will be different from those in Nevada and vice versa. Many times, I have encountered fire department representatives or building inspectors who have been incorrect in their assessments of a situation or a code. Do not be intimidated by these officials. If you are right and they issue you incorrect information, challenge them. Explain how you perceive the law or code and ask them to verify their information. With the constant changing of laws and codes, inspectors are often in error.

While many individuals will gain valuable information from this book, it is basically written for the security/FSD, in particular, the law enforcement or security professional who has no experience in the field of safety.

The book explains the functions of a fire safety department and how it applies to your facility. In no way is this book to be viewed as the overall authority regarding fire safety, but used properly, it will provide the reader with a better understanding of fire safety.

The author and publisher assume no responsibility or liability for any action taken by others as a result of the material presented in this book. Seek the expert assistance of competent, qualified professionals.

New and existing building owners will benefit greatly from the information and ideas discussed throughout the text.

Fire Safety and Loss Prevention

1

Human Behavior and Fire

In 1934, a fire at the Kerns Hotel in Lansing, Michigan, claimed 35 lives. In 1943, at the Gulf Hotel in Houston, Texas, 54 persons died in a fire. In 1946, two catastrophic hotel fires resulted in fatalities and destroyed property—the LaSalle Hotel fire in Chicago claimed 61 lives and the Winecoff Hotel fire in Atlanta claimed 119 lives. In the 1970s and 1980s more deaths and property destruction were caused by fire in major hotels. Fires at the Beverly Hills Supper Club (1977), the MGM Grand Hotel in Las Vegas, Nevada (1980), and the Westchase Hilton Hotel fire in Houston, Texas (1982) caused the fire safety, security, and insurance agencies to reevaluate the hotel industry's fire safety procedures and systems.

The disaster wrought by these hotel fires underscores the need for management to interface with security and fire safety officials, and to develop and implement strategic fire operation and safety plans. These plans must consider the element of human behavior during a fire. Developing, establishing, and implementing your fire operation plan, as well as training your staff, must allow for the unpredictable nature of human behavior.

Presently, we cannot predict what will happen during a fire. Individuals respond differently in crisis situations. As the director of security or loss prevention, you cannot foresee how certain persons will react during a fire. Therefore, human behavior is the *X-factor*—it cannot be put in a plan or formula when you establish a fire safety plan.

"There are at present no official strategies for the life safety that can stipulate a correct sequence of actions to follow . . . the physical environment cannot be designed to literally reflect a specified sequence of actions . . . no two fires are alike . . . "[1] This statement is the most popular statement made by fire fighters and researchers. Each building is structurally different. Therefore, determining how and when fires originate is difficult, if not impossible. Fire sources can range from an electrical problem in an apartment's fuse box to a boiler explosion in a high-rise hotel. The origination of every fire depends on many circumstances and the manner in which occupants react in the apartment building will be vastly different from the reactions of the hotel occupants.

MALE AND FEMALE BEHAVIOR PATTERNS

Human behavior is an expression of subconcious drives or tendencies, such as instincts, needs, constitutional makeup, or preprogrammed action patterns, that appear at different stages of maturation.

Understanding behavior patterns and traditional male/female roles displayed during a fire is important. Security/fire safety directors (FSDs) should consider this information when they design fire safety plans, such as preparing a plan for an all girls college. Females are more likely to try and warn others, while males will attempt to fight the fire. "Females are more likely to warn others and wait for further instructions (for example, if husband and wife are both present). Alternatively, they will close the door to the room of fire origin and leave the house. In both cases, females are more likely to seek assistance from neighbors. Male occupants are more likely to attempt to fight the fire. Male neighbors are more likely to search for people in smoke and attempt rescue."[2]

Another strong human behavior pattern is to seek verification of a fire. Both sexes want to be assured (through various cues) that a fire is in progress and they must evacuate.

A research study of human behavior during a fire is not conducted during a fire, but after a fire. Questionnaires are distributed days, sometimes weeks, after a fire is controlled. During a fire, human instinct is to survive. The responses and actions of the survivors are studied and explored afterwards.

The four following categories depict the behavioral differences in males and females during a fire. These categories are based on the attitudes and opinions of males and females who survived the Beverly Hills Supper Club fire on May 28, 1977.

1. *Fire Awareness*—The different responses from men and women were determined pertaining to the means of fire awareness, time of fire awareness, primary and secondary actions of the guests, and third and final actions of the guests.

2. *Attempted Evacuation and Evacuation of Guests*—Different responses were determined pertaining to how men and women attempted to leave the supper club and how they actually vacated the club. Forms of egress, such as windows, doors, stairways, and fire exit doors, were discussed with both sexes.

3. *Refuge Process*—Male and female actions were determined pertaining to those areas that were thought to be safe places for guests during the fire. Most guests converged in stairways, bathrooms, and the kitchen.

4. *Behavior and Actions of Guests*—Differences in male and female behavior were determined and opinions evaluated of persons who tried to evacuate or remained behind to fight the fire. Differences were

discovered in the numbers of males versus females who attempted to warn others of the fire as well as tried to obtain personal effects, such as coats, hats, and jewelry.

These categories can aid security directors in their quest to understand human behavior during a fire. However, note that the male/female decision-making process changes as the fire expands or diminishes. Beginning with "fire awareness" and concluding with "behavior and actions of guests," the decisions made during a fire change every few seconds. New decisions are constantly being made depending on the fire's location and a person's perception of what is happening. "Decision making during the various stages of a fire emergency can present a severe challenge to the participants. Since every situation is somewhat different, successful coping with a fire emergency can demand more of the participant than following a set of previously memorized correct actions."[3] A behavior pattern that develops during a crisis can also be considered a *norm*. A fear of the unknown, the fear of fire, the general fear that something ominous is about to happen makes our norms readjust to help us survive. During a fire, people focus on surviving; we formulate a plan to escape the fire. The myth of *panic* killing people in fires is false and generally exaggerated, as stated in "The Myth of Panic" by John P. Keating.

> Multiple deaths in fire tragedies are frequently headlined in the press by reports of panic behavior of the victims. Such conclusions by the press persist, in spite of the insurmountable research evidence that concludes exactly the opposite.[4]

Keating's premise is that newspapers and television use yellow journalism to sell their stories. He states that after reading these stories, he feels the word *panic* is grossly abused and certainly overused.

FSDs can apply Keating's theory when developing their fire operation plan. Why is so much emphasis placed on human behavior? It's simple! No two fires are alike and no two persons act alike during a fire. Fire safety and security directors must always remind themselves that it is impossible to predict behavior during a fire. Therefore, formulating a structured fire safety plan and training security and building maintenance staffs to act as fire safety personnel is the first step to minimize losses in the event of a fire. A reliable plan and rigorous training will make it easier for staff members to make appropriate decisions during a fire. Coordination, advanced planning, education, and retraining are four other criteria that help make a fire plan operational. Conflicts develop when there is no advanced planning.

Organization and planning are difficult to do during crisis periods. Imagine the chaos that will ensue if there is no fire safety plan, no designated organizers, and a fire is raging two floors below you. Mass confusion will develop, while persons run rampant throughout the facility seeking escape. You cannot factor this behavior into your fire operation plan, but you must consider it. The four following criteria must be met before any behavior in a fire is labeled as *panic*. Keating outlines these four criteria as follows:

1. "... the typical panic response manifests a hope to escape ... escape through dwindling resources."
2. "... there must be some kind of contagious behavior, especially if key-noted by leaders in the community that is effected by the fire."
3. "... each person has to be aggressively concerned about his own safety, as opposed to concern for other people who are also in the fire."
4. "... and most important, there must be irrational, illogical types of responses."[5]

Human behavior in a fire condition depends solely on the individual. Whether or not hypervigilant behavior is displayed in a given situation depends on the definition of the word *panic*. *Hypervigilant behavior* is an emotion; *panic* is an action with aggressive or irrational components. An individual's anxiety during a fire ranges from high to low, depending on their emotional assessment of the fire.

SUMMARY

At present, there are no valid methods to predict human behavior, the X-factor, during a fire. Individuals' primary objective during a fire is to escape to safety with their family members and, in some instances, their possessions.

Short and long-range planning are crucial to a security director or manager. However, when establishing a fire safety plan, we must not forget the X-factor. Human behavior, under any circumstances, cannot be incorporated into a fire plan. No two fires are alike and no two individuals act alike in a fire. Security managers and FSDs who fail to incorporate the X-factor in their plans are not being realistic regarding fire safety.

The male and female behavior patterns that have been studied and analyzed over the years are a good indication of how people will react during a fire or smoke condition. However, these behavior patterns change from fire to fire and from person to person. Males and females react differently from one another during a fire. However, while our instincts make us react differently during a crisis, both males and females seek to survive any crisis situation without being injured.

John P. Keating, a pioneer in the field of fire science, has aided the fire science field by researching, studying, and writing about human behavior. Keating describes four types of human behavior that are exhibited prior to panic taking over as the overriding human emotion. According to him, panic can be defined by the existence of the following: hope of escape, contagious behavior, primary concern for one's own safety, and most importantly, irrational types of responses. Keating also states that the myth of panic during fires is greatly exaggerated. However, FSDs must incorporate panic into their fire operation plan.

2

Fire Safety and Security

As the security director or loss prevention manager of a major retail chain, shrinkage and employee theft is a problem that you must deal with continuously. Your undercover operatives may inform you that most of the merchandise is being taken through an unsecured fire door. An easy solution to the theft problem is to simply chain and lock the fire door. This action will prevent thieves from exiting with the store's merchandise. However, due to the fire codes in all municipalities, fire doors must provide egress in the event of a fire. If you chain the door, you are negligent in conforming to the municipality's fire codes. If the fire department inspects your facility and finds that fire exit doors are chained, thereby preventing persons free and clear passage, you will be issued a hefty fine. Besides risking a hefty fine, you are also risking human life in the event of a fire. But what about the store's shrinkage crisis? As you will read later, there are other methods available to you when confronted with this situation. You must use all available resources to effectively deal with this situation.

As the security director of a hospital, you may be informed that, due to budgetary constraints, you have to cut back your staff. However, assume that you are faced with a problem of unauthorized persons entering the hospital after visiting hours through certain entrances. It is crucial that you physically monitor these entrances. How can you do this with reduced manpower? You have a variety of options that are suited to protect the hospital against unauthorized entry, but will they conform to the building codes and fire codes? If they don't, you must rethink your options or your actions will be costly to the hospital.

Let's assume that you are the security director at a university that presently has one of the best fire detection systems in the country (it conforms to the building and fire codes) and the forensic lab receives a $1,000,000 grant to add three new labs, two offices, and increase its staff. The college has already hired an outside architect to design the layout. When should you become involved in the design? Almost immediately, so that you and your staff can have a better understanding of the building than the architect. The architect will be wise to seek your assistance and input from the beginning of the design process. You won't design the layout, but you will suggest where certain doors, entrances, and exits should be placed. By strategically planning building layout prior to actual construction, you may save the university some money.

INTERFACING SECURITY AND FIRE SAFETY

The protection of life and property should always be placed first and foremost on a list of priorities. However, a company's bottom line does not always reflect this thinking. For example, hotels lose millions of dollars each year due to theft. Employees as well as guests pilfer towels, bathrobes, and other items daily. As a security director/FSD, it is your responsibility to protect the lives and property of hotel guests and employees, while management seeks to maintain high occupancy rates based on sales, affordability, location, services provided, and good fire safety and security methods. Oftentimes, management's goals conflict with security's goals. You need statistics to support your need for a fire safety plan and security staff. As the security/FSD, it is crucial for you to justify a fire safety strategy to management using statistics. Begin by interfacing fire safety and security. When you generate your monthly reports, present the statistics for both categories in one comprehensive list. An example of this is shown in Table 2–1.

The seven asterisked categories deal directly with fire prevention and fire safety. As the security/FSD, you are achieving two significant goals with this type of report. You are giving management the opportunity to view your department's monthly activities and you are justifying your budget and alerting management to the fact that both fire safety and security complement one another.

THE IDEAL VERSUS THE REALITY

Political Conflicts

As security/FSD, you want to use good security practices and conform to fire, life safety, and building codes. However, established codes are not enough to ensure the safety of people and property in the event of a fire. Conforming to local laws can and will tie your hands. Security procedures will often be compromised in order to comply with mandated codes. However, once a code or law is mandated, it must be followed. Municipalities are responsible for establishing and enforcing fire and building codes. As the security/FSD, you are responsible for ensuring your organization adheres to a municipality's codes. It is crucial that you remind your organization that fire and building regulations are mandated, and compliance is required by law, whereas most security regulations are not mandated.

Table 2-1 Monthly fire safety/security report

Incident Classification	Total for Month
Aided persons	14
Alleged arson*	1
Alleged auto thefts	0
Arrests on property	2
Assaults	1
Breaking and entering	2
Burglaries	5
Disorderly conduct	8
Drunk and disorderly persons	7
False alarms registered*	11
Facility vandalism	2
Family disputes	3
Fighting	4
Fire(s)*	2
Fire department on property*	13
Fire drills*	3
Harassment complaints	0
Open doors	5
Police officers on property	16
Smell-of-smoke reports*	4
Sprinkler system(s) discharged*	2
Vehicle vandalism	1

Politics plays a major role in establishing specific codes. Usually a crisis heightens public awareness, thereby forcing politicians to address a specific issue. Unfortunately, it is only after a major fire in which many persons are killed and much property is damaged that code violations and enforcement of codes are addressed. For example, take the fire that killed 87 persons at the hispanic social club in the Bronx, New York, on March 25th, 1990. One day after the catastrophe, 180 special officers were assigned to close illegal social clubs throughout the city. Two weeks later, due to budgetary restrictions, the task force was reduced to 120 officers. "New mandates were going to be put into law. Building owners would be arrested if they did not comply," cried New York City Mayor David Dinkins. What New York City did, as do most cities when faced with a catastrophe is minimize the effect of the local interest groups—in this case, those who are opposed to complying with strict fire safety codes.

Municipalities then formulate strategies to revise existing fire codes and standards in order to embrace current technology. Due to the lack of communication between fire and building inspectors, and the lack of interest among politicians to ensure the codes are enforced, conflicts will continue.

Financial Conflicts

It is expensive to comply with a municipality's fire and building codes. As the security/FSD, you must incorporate these costs into your budget. Inspect your facility now by conducting a comprehensive Risk Analysis Audit (RAA) for both security and fire safety. According to *The Factory Mutual Systems Approval Guide*, the purpose of an RAA is to elicit corporate self-evaluation. An RAA helps management develop sound attitudes toward risk exposure, recognize exposures to loss, and, if necessary, seek further clarification about risk from specialists in the loss prevention field.

The facility you are responsible for protecting will vary in design, population, and building classification from others for which you may have been responsible. Designing an effective RAA will not be easy. However, you must begin somewhere. As an example, imagine that you are responsible for conducting an RAA for a hotel that is less than 75 feet in height and has an occupancy that is considered transient, which means occupants stay less than 90 days. The hotel has a wet sprinkler system throughout the seven-story facility. As the security/FSD, your primary responsibility is the protection of life and property. You must design an operational fire safety plan. You can better fulfill the responsibilities of both jobs by completing a Fire Safety Audit. (See Figure 2–1.)

After you complete your fire safety audit, you must now begin taking measures to ensure that your plan is realistic and conforms to your budget. When you are responsible for both security and fire safety, you need every department's cooperation to create a safe and secure environment. You cannot assume that because you are aware of mandates and regulations that other department managers are aware of them as well. Many times, you may have information that other department managers see as a nuisance. You must be able to convince them that their cooperation is essential to conduct a thorough fire safety survey of the facility.

The amount of combustible material in a facility, including the combustible material within the building structure, determines the fire load. Knowing the contents of your facility allows you to incorporate the fire load into your overall plan. The strategy is to gather the appropriate information regarding the facility and then use it to properly secure and keep the facility safe from fire. Ask purchasing department or department managers if their furniture is constructed of a high percentage of noncombustible materials. Also inquire if the materials or woods used are fire retardant or treated. Inquire if the upholstery and plastics in the offices are self-extinguishing. Drapery linings and curtain materials should be fireproof. The carpeting and carpeting underlayment should have a low flame spread and smoke-developed ratings. Smoke-developed ratings measure how quickly a material emits smoke when burning.

Date: _____   Reviewer: _____

Hotel: _____   Address: _____

1. Are doors to individual guest rooms self-closing.	YES	NO
2. Are doors to stairwells closed and self-closing.	YES	NO
3. Are stairwells unobstructed and lettered?	YES	NO
4. Are doors that subdivide corridors self-closing?	YES	NO
5. Are exit lights operational?	YES	NO
6. Is emergency lighting operational?	YES	NO
7. Are standpipes properly colored and operational?	YES	NO
8. Are fire extinguishers properly positioned and charged?	YES	NO
9. Are interior alarm systems operational?	YES	NO
10. Are manual pull stations operational (i.e., glass rods in place)?	YES	NO
11. Is the fireman's telephone outlet operational?	YES	NO
12. Is the speaker paging system operational?	YES	NO
13. Are the smoke detector heads operational?	YES	NO
14. Is the sprinkler control valve seal open and labeled?	YES	NO
15. Is the floor plan posted near the elevator on each floor?	YES	NO
16. Is the base of the elevator shaft free of excess debris?	YES	NO
17. Are Detex watch clock stations in place and operational?	YES	NO

18. Are the fire safety plan and master keys available for the fire department? _____

19. List any fire safety infractions you observe during your inspection.

_____

Signature: _____

Figure 2–1   Fire safety audit sheet

Budgetary Conflicts

From a financial viewpoint, it is important to note that most security managers allocate very little money for fire safety and prevention. If your facility has two separate and distinct departments (i.e., the fire safety department and the security department), it is not cost-effective. Due to financial restraints, most organizations merge the two separate departments into one department that has a goal of protecting life and property.

Security directors are not only crime fighters and fire fighters, they are also business executives who accept full responsibility for the fiscal management of their departments. Through fiscal planning, security/FSDs develop future expenditure projections for staff, equipment, facilities, and programs to achieve department goals and objectives. Any projected budget is a director's statement that monies are required to launch, maintain, or expand programs, functions, and activities. Justification and need are the elements on which directors carefully plan their budgets. Management usually reviews these budgets, which have been substantiated with adequate justification of all major items. Due to the typical revenue limitations of most companies, FSDs must be persuasive, yet objective, with administration. Exhibit sound analyses with your budgetary projections in order to gain budget approval.

Security/FSDs are accountable for all aspects of their departments' fiscal policies, processes, and controls. At times, you may ask a subordinate to prepare a budget for a special program or a new division, however you are still responsible for reviewing it and must demonstrate sound judgment in its use.

Annual budgets should be developed in conjunction with all other departments in your organization. All departments weigh their actual needs versus their wants. Budget proposals should be reasonable and economically sound. Administration will have the final decision on all programs. If your proposed programs are presented professionally and are cost-effective, they have a chance of being accepted or placed on a tentative final budget.

Security/FSDs complain that they lack the financial support from administration and therefore cannot run a professional department. When administration reviews and rejects their departments' request for allocated funds, many times the director failed to either justify the budget or the budget was poorly written or researched. Detailed justification for a budgeted item should be written and researched professionally. Never assume that you will be provided with a budgetary request simply because you asked for it!

Cost-effectiveness is a term administration likes to hear. Whenever possible, demonstrate your department's cost-effectiveness. When a potential problem arises, identify it and act on it. It is crucial that every supervisor or manager participate in determining a department's budget. Sometimes it helps if the participation extends to the lowest level in the hierarchy. Getting your department involved creates a cost-effective atmosphere while exploring all possibilities.

Many security/fire safety departments initiate new programs and then fail to evaluate them, especially on the basis of cost-effectiveness. An energetic evaluation of programs should be conducted periodically to determine program success and costs.

Throughout the security/fire prevention industry, directors want to institute correct and sound programs, but do not have any notion of how to prepare a cost-effective budget. Remember that administration wants an analysis of the monies spent during the previous year and a projection of the budget you need for the following year. Do not be intimidated by budgets or budget preparation. Each company has a controller or a fiscal affairs officer who oversees all budgets. If questions arise or you are unsure of items within a budget, ask the controller for advice. You only hurt your department's chances of acquiring an accepted budget if you don't seek the advice of the controller or fiscal affairs officer. An example of a clear and reasonable budget request is detailed in Table 2–2. Notice it compares the prior year's

Table 2–2 Budget preparation proposal

	1990 Budget	1991 Request	Director's Recommendation
Personnel			
Director	$45,000	$47,250	$47,250
Managers	33,000	34,650	35,000
Supervisors	28,500	29,925	30,500
Peace Officers	24,000	25,200	25,200
Officers	21,500	22,575	22,575
Clerical	19,500	20,475	20,475
Overtime @ 8%	3,640	3,822	2,389
Part-time Help	5,750	6,500	7,250
Totals	**$177,250**	**$183,897**	**$190,639**
Expenses	15,000	17,000	17,000
Equipment	45,000	11,000	9,000
New Programs	8,000	6,250	8,000
Direct Totals	**$245,250**	**$218,147**	**$224,639**

expenses to project expenses, as well as shows a column for the director's recommendations.

Staff benefits such as insurance, retirement, and major medical coverage will be accounted for by the fiscal officer or controller. Remember that the numbers listed in your budget must be multiplied by the number of personnel in each job category. Your estimates must be based on current cost-of-living figures. Many times, the numbers will be projected over a 3-year period, if you are dealing with union shops.

Unforeseen conditions invariably arise when the need for additional security or fire safety services cannot be forecast by prior fiscal planning. Therefore, it is imperative that interaccount transfers be available to the security/FSD for such emergencies. Numerous modification mechanisms should be available within your budget to provide for these contingencies. Alternatives include transferring funds from a later funding period to the present period, transferring funds from an account that has a savings to one that requires additional funding, and asking for a stipend of supplementary capital for emergency conditions.

In many ways, security/FSDs run their departments like police or fire department administrators. They keep abreast of changes in technology and weigh their potential to assist in the security and fire safety fields. They achieve their goals through long-range planning and budgetary funding.

Planning is crucial and correct timing is essential to developing a security and fire safety department. Your administration must decide, at the outset, if they want two separate departments. A sound budget usually justifies why the organization should combine the two departments.

"Although most police agencies have neither the manpower nor the resources to undertake an in-depth research and development program, all agencies should encourage such programs by participating in them when possible and by taking advantage of the results."[6] This quote appeared in an article of The Police Chief in May 1970. Some twenty years later, it still applies to the security and fire safety industries; remember it when preparing budgets and justifications.

The Conflicts Begin

Crime prevention gauges should be an identifiable component of existing or proposed regulatory codes. Building, fire, and safety codes should be evaluated by regulatory administrations and private security professionals to avert confrontation when implementing practical crime prevention measures.

Presently, a myriad of codes and regulations exists throughout the fifty states. Different codes are present at the local, state, and national levels concerning fire, building, and life safety. These codes mandate what can and cannot be done in the security and fire safety fields. In many instances, these codes conflict with one another, and at other times they complement each other. For example, San Francisco currently adheres to four separate electrical codes relating to alarm systems.

"The latest development in regulations and codes is to draft building security codes. Some success in crime reduction has been documented where security codes have been implemented. However, it is preferable to incorporate crime prevention measures into existing building, fire and safety codes. Additional codes only create another bureaucratic burden for the businessman and the citizen. Further, it appears likely that the future direction of building codes is toward developing a single code covering building, fire, safety, and possibly other environmentally related topics,"[7] This idea was proposed 15 years ago. It is a sound idea, but has virtually no substance. Unless the regulating bodies choose to accept security standards on the same level as fire safety standards, this idea will remain a vision.

Throughout the United States, crime and loss prevention practices are constantly clashing with building, fire, and life safety codes. As most security professionals are aware, good security frequently exists contrary to fire regulations. An illustration of this dilemma occurred in New Orleans in January 1973. A gunman entered a Howard Johnson's motel via an exit door that connects the building's garage with the motel's upper story guest rooms. New Orleans' fire regulations required the door as an emergency exit from the garage. The door was equipped with a crash bar designed to allow access to the motel's fire stairwell. Because the crash bar allowed access to the stairwell, the sniper gained entry to the guests' rooms.

"The scope of security building codes also varies. Hollis DeVines, director of the Schlage Security Institute, points out two examples: In Indianapolis, Ind., single-family dwellings are included in security building codes; in Montgomery County, Md., the code is retroactive for motels, hotels, and multi-family dwellings. Other codes specify only commercial buildings. The National Institute for Law Enforcement and Criminal Justice's Federal Security Code has provisions that are so broad that they apply 'to all existing and future buildings or structures.' "[8]

SUMMARY

Conflicts are prevalent in the security and fire safety industries. Twenty years from now conflicts will still be present. Knowing that these conflicts exist should put security/FSDs' minds at ease in that they are not alone in facing this dilemma.

Determine the conflicts in your municipality and examine how they affect your facility. Be specific when researching your facility; understand the various codes and know who enforces them. Be prepared to act, rather than react, when dealing with codes and regulations.

The ideal versus the reality of political pressures and budget preparations and justifications is vital for the director to establish. Instituting awareness programs, fire safety programs, and loss prevention programs are part of the ideals that security/FSDs want their organizations to create and maintain. Of course, managers/FSDs must operate under budget restraints, and often their ideas must be compromised by financial realities. Justifying your expenses can go a long way in attaining your goals.

3

The Security and Fire Safety Department

As security/FSD it is imperative that you prepare your department to perform the functions of a security and fire safety department. Unless your organization already has a fire brigade or fire safety unit in place, it is your department's responsibility to act as fire safety officers when a crisis arises.

The composition and organization of your department will reflect your planning and budgeting as an effective security/FSD. Most organizations request that the security department become involved with any crisis situation and, most times, the director and his staff are faced with incidents beyond their control. However, fire safety is a major concern and there are mandates that require a knowledgeable individual to implement and direct the fire safety functions of the organization.

It is necessary for a security/FSD to have experience in effectively dealing with regulating municipalities. Usually, administration relies on the security director to handle summonses from the fire department. Administration seldom knows what is involved when fire safety regulations become a priority.

FIRE SAFETY DIRECTOR AND
DEPUTY FIRE SAFETY DIRECTOR QUALIFICATIONS

The qualifications of the individuals who become FSDs and deputy fire safety directors (DFSDs) are crucial and meeting them is mandatory in some states.

As a general guide, I have listed seven important criteria that both the FSD and DFSD should meet in order to qualify for either one of the positions.

1. The persons shall possess at least 5 years experience in the fire protection and fire prevention fields, have 5 years experience in a responsible

defined in the city's building code, or a satisfactory combination of equivalent experience.

2. The persons shall satisfactorily complete a course for the position of Fire Safety Director, given by an accredited school or organization, that is acceptable to the city's fire department.

3. The persons shall receive a passing grade on a written examination conducted by the city's fire department covering material provided in the course.

4. The persons shall receive a passing grade by the city's fire department regarding the characteristics and occupancy of the building that is to be under the applicant's control.

5. The persons shall be at least 18 years of age.

6. The persons shall have a reasonable understanding of the English language and be able to answer satisfactorily those questions that may be asked of them during their on-site examinations.

7. The persons shall produce evidence of their character and past employment that is satisfactory to the fire commissioner of that city.

In addition to these criteria, selection of an FSD and DFSD should be predicated on their ability to

be available for each tour of duty—7 days a week

recognize personnel, equipment, fire systems, and other provisions of the organization's fire safety plan

deal effectively with high-stress situations and recognize potential fire hazards

comprehend written and verbal instructions, and delegate authority when required

relay written and verbal instructions to the staff

prepare concise, detailed, written reports

effectively communicate instructions to the staff during a crisis and noncrisis

prepare budgets that correlate to the functions of the department

These guidelines should be followed by any organization that wishes to hire individuals for a fire safety program. It is imperative that an FSD and DFSD be qualified according to the guidelines established by the fire commissioner of the city. It is a waste of time and money to select a candidate that is unqualified to act as an FSD or DFSD.

CHOOSING QUALIFIED CANDIDATES—THE FIRE BRIGADE

After a determination is made of qualified FSD and DFSD candidates, you will then select employees from your present staff to become members of your organization's fire brigade. The selection process, again, relies on compliance with your city's fire department. Fire brigade personnel should

be physically fit and able to perform under stressful situations

be familiar with other personnel and equipment, and have a working knowledge of the organization's fire safety plan

be able to comprehend and implement written and verbal instructions, and apply them to crisis situations; sometimes knowledge of a second language is required

be able to communicate effectively during a crisis

After you have determined the employees who are qualified to become members of the fire brigade, implement a program to ensure that the fire brigade is compliant with the organization's fire safety plan.

In choosing qualified fire brigade candidates, select employees who are thoroughly familiar with the building's layout, occupancy, and street alarm box location, and familiar with fire-fighting techniques. Select those fire brigade members who have satisfactorily completed a fire training program developed and administered by the FSD. Fire brigade candidates must also pass a written examination, administered by the FSD, that covers material on the organization's fire training program.

When developing procedures for your fire brigade to follow, you must consider the classification assigned to your building by the city's regulatory bodies. A city's regulating bodies assign classifications to various types of buildings. Hotels, schools, hospitals, retail establishments, industrial plants, and others are assigned a designated building code to conform with city ordinances. Many of these buildings are then issued a certificate of fitness from the city's fire department. "Certificates of fitness are issued to individuals or businesses with demonstrated proficiency in skills, training, and testing in areas that affect fire safety. This includes pyrotechnics and persons who handle explosives; persons who install fire protection equipment and systems, including sprinklers; and persons who perform maintenance on fire extinguishers and fire detection systems."[9] In some municipalities, certification specifications may include a financial bond or liability insurance. You should verify this with your organization's insurance carrier.

When you begin formulating your organization's fire operation plan, many obstacles arise. The aspects of security and loss prevention seem trivial when compared to what you must accomplish in order to comply with your city's mandates regarding fire safety and prevention. However, never lose sight of the fact that your department must also continue to operate toward reducing shrinkage and preventing theft.

A HISTORY OF FIRE BRIGADES AND FIRE TRAINING

The first organized fire protection unit was established when Augustus became ruler of Rome in 24 B.C. Augustus had the foresight to create a watch guard service to look for fires and prevent them from starting. The fear of a fire's capacity to cause death and destruction was just as prevalent then as it is today.

In 872 A.D., one of the earliest recorded fire protection ordinances was introduced in Oxford, England, when a curfew was adopted requesting that hearth fires be extinguished at a certain hour. The earliest known, organized fire brigades were called fire insurance brigades. They were established in England, in 1666, as a result of the great London fire. Prior to that, in 1643, during the British Civil War, women were organized to patrol the town of Nottingham during the night and to put out fires and prevent new fires from starting.

It was not until Edinburgh's 1824 Fire Brigade Establishment that public fire services began to develop modern standards of operation when a surveyor named James Braidwood was appointed chief of the brigade. He selected 80 part time aides between the ages of 17 and 25, and required regular drills and night training. Braidwood wrote the first comprehensive hand book on fire department operation in 1820.[10]

Up until 1853, all fire departments had volunteer workers. Most cities had no training programs, lacked discipline, and had no positive direction. Firefighting was not a paying job and the work was hazardous. On April 1, 1853, in Cincinnati, Ohio, the first paid fire department was established. The department's only equipment to combat fires was manpower and horse-drawn steam pumpers.

An interesting event occurred in 1855 when two steamers were delivered to New York City to aid the volunteer fire department. The volunteer fire department refused to use them. Controversy and criticism continued to plague the New York City Volunteer Fire Department for 10 years for not using the equipment offered to them. Then, in 1865, the Metropolitan Fire Department, using steamers, officially replaced New York City's volunteer fire department.

At this time, other cities throughout the country began to see the importance of a trained and equipped fire department. It was imperative that a city's fire department be prepared to deal with any fire emergency. The first fire drill school, at which basic fire training and company drills were performed, was established in Boston, Massachusetts, in 1889. New York City established the first fire college for advanced fire officer training in 1914. Also in 1914, North Carolina established that state's first school. In 1937,

Oklahoma A&M College (now Oklahoma State University) instituted a 2-year college program in fire protection. By the 1950s, most states were providing methodical fire service training. Most cities were eager to join in a fire training program. The 1960s, 1970s, and 1980s saw fire department organization and planning grow by leaps and bounds. Industry was becoming more aware of the need for strict regulations in the fire safety field. In the past 30 years, fire department objectives and new legislation outweighed corporations' needs. Fire safety, training, and prevention has now become an integral part of today's business operations.

YOUR ORGANIZATION'S FIRE BRIGADE

After the FSD has been approved by administration and accepted by the municipality's governing body, it is the responsibility of the FSD to focus on two important duties—design and implement the organization's fire safety plan, and design and implement the organization's fire operation plan. The difference between the two plans is simple. The fire safety plan, in most municipalities, must be submitted to the city's fire department for approval. It is an overview of the organization's fire operation plan. The fire operation plan remains at the facility and outlines the functions of the fire brigade. These duties should be specific to your facility. Members of the fire brigade should include the FSD and DFSD, the alarm box runner and the backup alarm box runner, the fire guard, the fire fighter, the evacuation officer, the direction officer, and the communication officer. The duties of the fire brigade will be discussed in Chapter 4.

The FSD should design an operational flow chart and post it in the security/fire prevention department. This chart will detail the breakdown of the fire brigade. (See Figure 3–1.)

If your organization *does not* have a fire operation plan or a fire safety plan, would you know what to do if a fire began? If your facility *does* have a fire operation plan, when was the last time you saw it? Has the fire operation plan been updated in the last 6 months? Do you know where the nearest fire alarm pull box is located outside your facility? Negative answers to these questions indicate that there may be a potential fire safety problem at your facility.

PROPRIETARY OR CONTRACT GUARDS— HIRING AND TRAINING

Proprietary Guard Hiring Procedures

For the past 15 years, an argument has existed as to whether to use proprietary (in-house) or contract agency guards. Your administration may ask

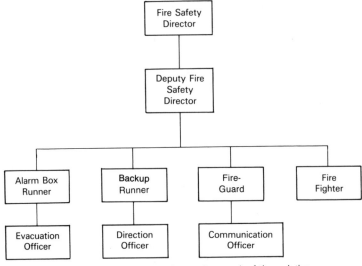

Your staff should be trained to perform each of these duties.

Figure 3-1   Fire operation flow chart.

you which alternative is the most beneficial to your operation. Base your answers on fact and use a budget to substantiate your decision.

Assume you are the director of a proprietary staff and administration gives you the responsibility of selecting and training a fire safety department. Where would you begin? If your organization is large and has a personnel department, your first step is to coordinate the hiring process with the personnel department. However, do not leave ultimate hiring authority to the personnel department. Meet with members of this department and advise them of the qualifications you are seeking in recruiting fire safety officers. Use the detailed criteria previously listed in this chapter. Many personnel departments do not know the standards a fire safety director uses to select a fire brigade candidate. It is an education process for them, as well as for you. Their function is to screen qualified personnel, conduct background checks, and verify job experience and education.

Today, more and more colleges and universities are offering major areas of study in security management and fire safety. Various professional organizations are also issuing certificates from qualified and accredited schools and it is imperative that you and the personnel department realize that professional certification programs do exist. For the past 10 years, the two most popular certifications in the security field have been the Certified Protection Professional (CPP) and the Certified Protection Officer (CPO) programs.

In an effort to recruit and select qualified fire safety applicants, the National Fire Protection Association (NFPA) has developed standards established by the National Professional Qualifications Board regarding standards, performance, and objectives. Some standards with which you and your personnel department should be familiar are

NFPA 1001, Standard for Fire Fighter Professional Qualifications

NFPA 1002, Standard for Fire Apparatus Driver/Operator Professional Qualifications

NFPA 1003, Standard for Airport Fire Fighter Professional Qualifications

NFPA 1004, Standard on Fire Fighter Medical Technicians Professional Qualifications

NFPA 1021, Standard for Fire Officer Professional Qualifications

NFPA 1031, Standard for Professional Qualifications for Fire Inspector, Fire Investigator, and Fire Prevention Education Officer

NFPA 1041, Standard for Fire Service Instructor Professional Qualifications

Colleges and universities offer these certification programs, which inferface with their municipality's fire departments. Candidates can receive certificates for completing 6-to-10-week courses in any of the previously listed categories. Certificates are awarded to the individual who passes a written exam and meets the criteria of that city's fire commissioner.

How does your personnel department, or you for that matter, verify the applicant's background? Preemployment screening is an area that gives personnel departments cause for concern. The Employee Protection Act of 1988 made it unlawful for private sector employers to require that job applicants or employees submit to polygraph examinations. The law mandates that if employees or prospective applicants initially agree to submit to a polygraph examination, they can terminate the test at any time. Employers who violate the law can be fined up to $10,000 for each violation. Also, employers can be ordered to hire, promote, or reinstate individuals, because of the employers' violations. Individuals who bring civil suits under this law and win, can have their legal fees paid for by the employer.

Integrity tests, also known as *pencil-and-paper tests* have replaced the polygraph examinations. Preemployment screening is a thriving business, as most companies want to thoroughly screen applicants prior to hiring them. If you opt to use a company that does preemployment screening, the company you select should be reliable and their testing methods and results valid.

Does your organization fingerprint or conduct criminal conviction researches for all job applicants? Many organizations claim they do. If a criminal conviction check *is* made, who makes it and who receives the information?

Negligent hiring is another concern of the personnel department. If a background check is not conducted and your organization does not verify the credentials of a potential applicant, but hires the applicant who then commits a negligent act while under your employment, your organization can be sued for negligent hiring. An interesting case regarding negligent hiring practices was the *K-Mart Corporation and Nathan Christian* (K-Mart's Loss Prevention District Manager) *versus Minerva Martinez, Elsa Nichols, and Martha Mata* case. This case was resolved in the Court of Appeals, in Corpus Christi, Texas, in November 1988.

The three plaintiffs sued K-Mart Corporation and Nathan Christian for actual and punitive damages for slander, assault, false imprisonment, and negligent hiring. At the trial, the jury found K-Mart and Christian jointly and severely liable for their actions.

> The plaintiffs presented two different theories of recovery of punitive damages against K-Mart. Under the first, an employer can be vicariously liable for punitive damages for an employee's misconduct if one of the following is shown:
> (1) the employer authorized the doing and the manner of the act;
> (2) the employee was unfit and the employer was reckless in employing him;
> (3) the employee was employed in a managerial capacity and was acting in the scope of employment; or
> (4) the employer or a manager ratified or approved the act.[11]

The jury established that K-Mart sanctioned the acts of Christian and K-Mart was reckless in employing him. The plaintiffs were entitled recovery for Christian's reckless and malicious conduct of slander and false imprisonment under the pretext of vicarious liability.

Employers must be cautious to bypass the selection or retention of any employee who he knows or should know is inappropriate to deal with the individuals invited to a location by the employer. Since the courts have ruled that employers can be held responsible for the actions of their employees, hiring trustworthy and competent employees is essential.

The advantages and disadvantages of proprietary versus contract guards is widely debated. However, some advantages to having a proprietary guard force include the following:

1. You select your own personnel.
2. Your in-house training procedures and manual are written and enforced by you.
3. The organization's image is upheld by your procedures.
4. The loyalty and motivation of the in-house officers is generally positive and reinforced.
5. The in-house officer has a tendency to remain with the organization longer due to better pay and benefits.
6. The communication between you and your staff is greater than with contract guards as well as your communication with administration.

7. The possibilities of promotions in the department and throughout the organization are greater than with contract guards.
8. In-house officers are better supervised by higher paid officers who make the job their careers.
9. The overall knowledge held by in-house officers regarding the facility and its occupants is greater than with contract guards.
10. Local law enforcement and fire departments will become more familiar with in-house employees, thereby creating a low-conflict liaison.

Disadvantages to a proprietary staff include

1. Proprietary staffs sometimes have strong unions that side with brother unions and refuse to cross picket lines in the event of a strike.
2. Proprietary staffs become too familiar with other employees, thus rendering them less effective in enforcing company policy.
3. The overall proprietary salary, including fringe benefits and insurance, can be astronomical.
4. Excessive overtime costs in the event of a catastrophe or contagious illness could create havoc with the director's budget.
5. The FSD's staff can feel intimidated by the facility's administration and give them vital information prior to your knowledge.

Proprietary Guard Training Procedures

As the security/FSD, you will realize, very early, the importance of qualified, trained personnel. All departments use various ways of training employees. However, your employees must be trained as a jack-of-all-trades as well as a master-of-all-trades!

For the past 15 years, the security industry has failed to implement uniform background checks and policy checks, and has not established standardized criteria for training a security officer. Each security organization has its own training procedures or none at all. Traditionally, most officers receive on-the-job training. Now try to apply this training method to the fire safety department and you'll realize that officers must be formally trained and, in some municipalities, certified to become fire safety officers.

The one major advantage of training a proprietary staff is that the same employees will follow the same rules and regulations daily. When these regulations change, a directive can be issued that explains the new or upgraded policy. Employees will be advised of the change and the directive will be enforced almost immediately.

When training your proprietary staff in fire safety, remember that fire directives are mandated. The training your staff receives is based on your facility's classification, occupancy, and building codes.

As FSD, the goals and objectives you set are predicated by the amount of professionalism you want to incorporate in your department. You must think as a fire department chief thinks when he trains and educates his battalion. The goal of any fire department training program should be the provision of the best possible training so that each person within the department will operate at acceptable performance levels relative to their rank and assignment. Ideally, training courses should have their own instructional objectives, a list of enabling objectives showing how the instructional objectives can be reached, and stated methods that explain how anticipated or desired behavioral changes can be measured.[12]

The overall training program you establish and implement will be viewed and analyzed by your organization. The administration will be unfamiliar with instructional objectives that are mandated and they may become part of the problem. Your response is to either tell them, "that's the way the fire department wants it," at which point your credibility is zero, or you can include the administration in helping you find ways to meet your training objectives. By using the latter alternative, you are accomplishing two significant things—you are warning them of a potential problem that may need to be addressed and you are ensuring that they fully accept the fact that the security department will also act as the fire safety department.

Today, more security and fire safety departments are participating with local fire departments in joint fire drills and joint evacuation procedures. Even more of these efforts need to be established. They are beneficial to you, your fire brigade, and the fire department. This interaction accomplishes three specific goals:

1. The fire department becomes familiar with the layout and design of your facility.
2. The fire department becomes familiar with your fire operation plan and fire brigade.
3. Your organization gains the respect and confidence of the fire department.

The proprietary trained staff has some advantages over the contract agency trained guards regarding fire safety training. Why? Due to lower personnel turnover, the proprietary guard is more apt to actively participate in fire safety training.

Contract Guard Hiring Procedures

As a security/FSD, your responsibilities begin to diminish when your organization chooses to use a contract agency. Be aware of the three major sales tactics that representatives use to sell their services:

1.   All of our officers are properly trained.
2.   Thorough background checks are conducted on each officer.
3.   We have 24-hour supervision.

Let's examine these features more closely. As the security/FSD, you need to know how the officers are trained and the duration of the training. Secondly, you need to know the type of facility (e.g., a retail establishment or a shopping mall) the officer was trained to protect. Also, remember that you want officers who are trained in fire safety. How can you be assured that the agency will hire individuals who have a solid background in fire safety? Set specific standards you want met regarding those individuals who will be sent to guard your facility. Having the agency train their individuals in fire safety is beneficial to both of you. Incorporate the training into the hourly rate. This way there can be no misunderstanding later.

If your facility is a retail establishment that has 60 stores throughout five states, supplying good security coverage becomes difficult, but is not impossible. By using contract agency guards, it is the agency's responsibility to train and supervise these officers in loss prevention and fire safety. You must establish uniform loss prevention and fire safety programs, and be guaranteed that these programs will be implemented by the security agency. Your objective is to work within the parameters established and defined by your organization to provide the most effective security and fire safety programs.

Some states require a criminal background check be made of the contract guard. This is usually done by fingerprinting the contract guard. This is also an expensive undertaking, and in comparison, the proprietary officer is *usually* not subjected to this procedure. The fingerprints are submitted to the state, which conducts a statewide search for arrests and convictions pertaining to the fingerprinted individual. This search can take up to 10 weeks to complete and if the individual was convicted of a crime in Tennessee and is applying for a job in New York, his prints will not be on file in New York!

Most contract guard agencies use the pencil-and-paper test for prospective employees. They will then conduct a residence verification check, Department of Motor Vehicles check, and employment verification check. It is up to you to inquire which preemployment screening methods are used by the contract guard agency. You need assurance that the agency uses effective measures to screen all perspective job applicants. The agency is responsible for having all employment applications, I-9 forms, and W-4 forms filled out correctly.

The question you should ask regarding 24-hour supervision is a simple one—In the event of an emergency, how soon can I expect a supervisor to arrive at my facility? If you receive a blank stare, you have your answer. If the agency replies, "Anywhere from 20 to 40 minutes," that answer is reasonable.

Many residential and commercial buildings use more than one guard per shift. If your facility is budgeted for 336 hours of guard coverage, who supervises them? Many companies use in-house supervisors and agency guards to protect their facility. Let's not lose sight of the fact that a supervisor must be well-trained in fire safety and knowledgeable of your facility and fire operation plan. Employing a supervisor who does not know what to do at your facility during a crisis is a waste of time, money, and possibly lives!

Some advantages of a contract guard service are

1. The cost could be as much as 25% less than the proprietary staff in salary, alone.
2. Fringe benefits such as health insurance, pension, social security, and vacation are paid by the agency.
3. Hiring and preemployment screening is conducted and monitored by the agency.
4. Training and supervision is supplied by the agency.
5. All overtime is paid by the agency, unless the client specifically requests it or it is written into the contract.
6. If the facility is a union shop, it is the agency's responsibility to deal with the union regarding employment matters.
7. The process of retaining quality agency guards is left to the discretion of the security director.
8. An account manager or operations manager is usually assigned to oversee the account at no cost to the client.
9. All administrative costs are assumed by the agency.
10. Insurance and liability costs are assumed by the agency.

Some disadvantages include

1. The low wages create high employee turnover.
2. Employees' loyalties are split between the agency and the client.
3. The agency's screening and training standards can become lax due to lack of concern on behalf of the agency's supervision.
4. Constant reassignment of personnel may cause their top officers assigned to your account to transfer somewhere else.
5. Insurance costs can increase or be canceled during a contract; this cost cannot be transferred to the client unless the client requests it.

Remember that the protection of life and property is your first and foremost objective. While you may inherit either a proprietary or contract guard force, it is your responsibility to ensure that they are adequately trained and supervised in fire safety. How this is accomplished is determined by your ability to convince your administration that security and fire safety should interface.

Contract Guard Training Procedures

It is the responsibility of the contract agency to train all employees in security, loss prevention, the power of arrest, patrol procedures, and, sometimes, fire safety. However, this training is not mandated. Whether or not an agency chooses to train their employees, is their perogative. However, it is your responsibility, as the director of security, to ensure that contract guards are trained in the field of fire safety.

Studies have revealed that training for contract guards is limited or non-existent. "For example, a survey of members of the American Society for Industrial Security revealed that only 68 percent of the respondents provided formal training for new employees and only 48 percent required annual formal training."[13]

You must establish and implement fire safety training and procedures. Your fire operation plan must be designed so that contract guards can carry it out. Meetings with the contract agency's operations manager and dispatchers are essential. They will benefit you, as well as them, when implementing your fire operation plan. Try to establish a rapport with the contract agency and split the cost of training at your facility. Establish a 32-hour training course in security and fire safety for which the agency pays for 16 hours of training and your organization pays for the other 16 hours.

The security/fire prevention industry is an effective complement to law enforcement agencies and fire departments. This balance is accomplished through training, education, and professionalism. Even the most qualified and well-trained employee is *ineffective* if equipment and working conditions are not favorable to productive job performance. For example, your fire brigade arrives at the floor of a fire and their radios are not working, or the fire pull box has malfunctioned. Their training and knowledge cannot overcome faulty equipment.

FIRE PREVENTION

As the FSD, you must inform your administration that in order to comply with the fire department's mandates, *fire prevention* must be taken seriously. As the security/FSD, you must continually stress fire prevention and public safety. The NFPA has formed subcommittees, in the area of fire safety, for facilities that are affected by fire safety mandates. The facility you are responsible for protecting is covered by the NFPA's life safety codes in one way or another. These subcommittees are

Fire Protection Features

Building Service and Fire Protection Equipment

Assembly and Educational Occupancies

Health Care Occupancies

Detention and Correctional Occupancies

Residential Occupancies

Board and Care Facilities

Mercantile and Business Occupancies

Industrial, Storage, and Miscellaneous Occupancies

These subcommittees recommend and review changes in the *Code for Safety to Life from Fire in Buildings and Structures.*

Emphasis must also be placed on how a city's fire departments deal with fire prevention. "In Los Angeles and most large cities, the fire department or its fire prevention bureau has the power to do everything necessary to make the city safe from fire, for even without such enabling clauses in the charter or in the ordinances, such power would exist as a reasonable exercise of the police power, the latter being inherent in every state and its legal subdivisions."[14]

As the security/FSD, you must keep policing in mind when speaking of fire prevention. Your facility must be continuously patrolled with fire safety officers who look for fire hazards. You have the difficult task of educating personnel in fire prevention. Also, after a city ordinance is written or adopted pertaining to a fire prevention code, it is your responsibility to make sure this ordinance is upheld.

Experts in the field of fire science feel that "using a model fire prevention code is preferable to using one developed locally—because nationally developed consensus codes are based on a broad spectrum of fire prevention experience, they may protect a fire prevention official from being accused of developing a fire code that is too stringent or biased."[15]

The fire departments in your city conduct fire prevention inspections that are usually mandated. Inspected facilities include assembly and educational, public assembly, residential (excluding the interior of the property), mercantile, manufacturing, storage, and board and care occupancies.

Your city may differ from another regarding who and how these fire prevention inspections are conducted. In most cases, an officer from the bureau of fire prevention will conduct the inspection. However, due to budget and staffing restrictions, many fire departments are having their local fire companies or contracted representatives conduct these inspections.

The primary objective of these inspections is to ensure satisfactory life safety conditions within a specified occupancy. To accomplish these goals, refer to Chapter 2, Figure 2–1 which depicts a detailed fire safety audit checklist. All interior lighting, doors, stairwells, and fire detection systems must be in working order. If the fire prevention inspector finds that your organization is not in compliance with life safety conditions, you will be issued a warning or notice of violation.

Test your fire protection system and communication systems on a monthly basis to ensure you are prepared for surprise inspections. Regular testing should be conducted as part of your fire operation plan. This information should be logged in a separate fire inspection book designated solely for documenting the testing of the facility's fire safety systems.

If your smoke detectors are hard-wired or battery-operated, they, too, should be inspected and tested. Fire escapes, standpipes, sprinkler systems, and portable fire extinguishers must also be inspected.

Inspections do two important things—they educate the owners and occupants of the facility in regard to fire safe behavior and they ensure that life safety conditions are being met. As the FSD, you should welcome inspections. If your facility is fire safe and meets all the requirements of the fire department, no warning or notice of violation will be issued. If the fire department does find a violation, it may be something that you or your staff overlooked and you should be appreciative when it is brought to your attention. Based on her own judgment, the fire inspector will designate an appropriate period of time during which the violation must be corrected. The security/FSD or owner of the building must sign the violation and keep a copy. The organization then has a specified number of days to correct the deficiency before the fire department reinspects your facility. If the deficiency is not corrected within the allotted time, legal action will be taken against the FSD.

Due to myriad of new and old building and fire codes, fire protection is complicated. Professionals should be consulted when you are unsure of a regulation or face a crisis. A fire protection engineer (FPE) is one individual you may want to consult. An FPE is educated and trained in the technical areas of planning, education, and design. FPEs can easily communicate with architects, building engineers, and decorators to ensure that your facility meets specified fire and building codes.

The NFPA publishes the *NFPA 1, Fire Prevention Code,* that assimilates many NFPA principles by reference. Three building code organizations have also developed their own archetype fire prevention codes—the Building Officials and Code Administration International (BOCA) publishes the *Basic Fire Prevention Code,* the International Conference of Building Officials (ICBO) publishes the *Uniform Fire Code,* and the Southern Building Code Congress International (SBCCI) produces the *Standard Fire Prevention Code.*

These organizations publish their model fire prevention codes so that municipalities may reference them. The organizations take building design, electrical wiring, mechanical design, and plumbing and heating systems into consideration. Their codes are used to better protect a facility against fire and to promote code uniformity throughout the United States. "Each code references codes, standards and other official NFPA documents. These documents constitute the 'National Fire Codes,' considered the most authoritative set of fire safety regulations in the United States."[16]

SUMMARY

As a security professional working in law enforcement, you have probably had very little dealings with fire safety, fire prevention, and the many codes associated with maintaining a fire safe building. As you can see, this will change. A security department that also is responsible for fire safety duties, does so due to liability and budget constraints. Sometimes, in many large office buildings the engineering department will act as the fire brigade.

Selection of an FSD and DFSD is a difficult process. Candidates must be carefully chosen and their qualifications must be accepted by the fire commissioner of that city. When this step is completed, the fire brigade can be selected and trained. The fire brigade must be familiar with the facility and well-trained in the duties they are to perform. Training and educating are crucial to an effective fire brigade. Be aware that either proprietary or contract guards will comprise your fire brigade.

Fire safety and fire prevention are a security/FSD's primary objectives. Your are responsible for selecting and training fire officers and formulating fire safety and fire operation plans that are operational 24 hours a day.

4

In the Event of a Fire: Duties of the Fire Brigade, Fire Safety Director, and Deputy

In the event of a fire, the FSD or DFSD should immediately report to the designated fire command station. You must ensure that the fire department has been notified of a fire or smoke condition. This can be accomplished by telephone, if the phone lines are operational, or by sending the alarm box runner to the nearest street alarm box. You must also ensure that the interior fire alarm system has been activated. If your organization has an interior communication system, an announcement must be made to inform personnel that there is a fire/smoke or alarm condition in the building. Advise persons not to panic. Announce the exact location of the existing condition and instruct the persons within the facility to do the following:

Remain in a room or area, with the door closed, unless otherwise instructed.

Do not use elevators. Indicate stairways that can be used.

Do not return to the fire area for *any reason*.

Commence building evacuation. Provide routes and destinations, if possible. Instruct individuals to close all doors upon exiting their rooms or offices.

The above procedures are based on the assumption that the local fire department has not yet arrived to the facility during the beginning stages of the fire. When the fire department arrives, ensure the facility's perimeter is accessible and free for passage.

Ensure that your fire brigade has arrived either at the fire floor or the floor *below* the fire floor. Obtain the following information from the fire brigade:

the exact location of the fire/smoke condition

the severity of the fire/smoke condition

the number of persons trapped on the fire floor

the status of the evacuation (i.e., in progress or completed)

the status of the fire/smoke condition (i.e., extinguishable or containable)

any special problems encountered by the fire brigade

As the FSD, your job is far from over. You are the quarterback calling the crucial plays until the fire department arrives. Your decisions and judgments at this juncture may save or lose lives. While maintaining radio contact with members of the fire brigade, dispatch a fireguard to the floor *above* the fire. Obtain the following information from the fireguard who has reported to the floor *above* the fire:

severity of the fire/smoke condition

areas affected by the fire/smoke condition

status of the evacuation (i.e., in progress or completed)

number of persons trapped *above* the fire floor

any special problems encountered

Upon the arrival of the fire department, provide the senior fire officer with the following:

an updated copy of the fire safety plan

all sets of master keys

information concerning the exact location and severity of the fire/smoke condition

Then proceed to

explain the evacuation status (i.e., in progress or completed)

advise the fire officer of any persons who are trapped

advise the fire officer of any special problems that have been encountered

advise the fire officer of the locations of your fire brigade, such as above and below the fire floor

direct the fire officer and his company to the fire stair providing access to the fire floor

remain at the fire command station

Upon the arrival of the fire department, your job as quarterback is somewhat downplayed. The senior fire officer now has the responsibility of controlling the situation. At the direction of the senior fire officer, begin deploying members of your fire brigade to attend to crowd control and perimeter security duties.

Remain at the fire command station until the senior fire officer declares that the fire is under control and the fire department departs the facility. Facility administrators should be kept abreast of the fire condition. Do not give them a progress report every 10 minutes, but, at appropriate intervals, notify them of the fire/smoke condition.

After the fire/smoke condition is over, you must begin the crucial task of preparing written reports of the incident. Many times, it's best to inform administration that a full report will be forwarded to them within 24 hours—after you have investigated the incident and spoken with your staff. Sometimes it will take days, even weeks, before you submit a finalized report.

Review and critique the procedures and performance of those who participated in the fire operation. The objective is to evaluate the effectiveness of the procedures and the performance of those who participated. Post-fire analysis is discussed in Chapter 13.

MEMBERS OF THE FIRE BRIGADE

The fire brigade should consist of an alarm box runner, backup runner, fireguard, fire fighter, evacuation officer, direction officer, and communication officer.

In some commercial buildings in which there is no safety or security department, the building manager will assign individuals from various offices to act as floor fire wardens. Their responsibilities are similar, if not the same, as members of the fire brigade.

Alarm Box Runner

At the direction of the FSD or the DFSD, the alarm box runner should proceed immediately to the nearest street fire alarm box. It is important that this individual keep in communication with the fire command station. The alarm box runner, after pulling the fire alarm, should remain at that location until the fire department arrives. The runner should then direct the fire department to the location affected by the fire and only return to the facility at the direction of the FSD. Upon returning to the facility, the runner should assist in crowd control and perimeter security.

Backup Alarm Box Runner

The backup alarm box runner shall perform the same duties as the alarm box runner in the absence of or a miscommunication with the original alarm box runner.

Fireguard

The FSD should have fire guards on duty 24 hours a day. If your city's municipalities certify fireguards, ensure that your fireguards are certified. A fireguard patrols the facility with a two-way radio and a bull horn. In some municipalities, a guard is required for each 75 feet of the premises.

Fireguards must be able to recognize different classifications of fire and be familiar with the proper fire extinguishing agents used for each type of fire. There are four classifications of fire and specific extinguishers are recommended to combat each type effectively.

1. *Class A fires* are fires caused by ordinary combustible materials (such as wood, paper, and cloth), for which the quenching-cooling effect of quantities of water or solutions containing large percentages of water is most effective in reducing the temperature of the burning material below its ignition temperature.
2. *Class B fires* are fires caused by flammable petroleum products or other flammable liquids, greases, etc., for which the blanketing-smothering effect of oxygen-excluding media is most effective.
3. *Class C fires* are fires involving electrical equipment and the electrical nonconductivity of the extinguishing media is of first importance.
4. *Class D fires* are fires caused by ignitable metals, such as magnesium, titanium, and metallic sodium, or metals that are combustible under certain conditions, such as calcium, zinc, and aluminum. Generally, water should not be used to extinguish these fires. (See Figure 4–1)

Provide and maintain a log book for recording the fireguards' tour hours. Give fireguards clear and concise instructions regarding protocol in the event of a fire or a fire hazard. During fireguards' patrols, they should look for frayed and loose electrical wiring. They must also be sure that fire exits are unobstructed and unlocked at all times. In the event of a fire, the fireguard must

sound the nearest alarm using the manual pull station

report on the fire/smoke condition to the dispatcher using the two-way radio

ensure that the dispatcher acknowledges the report

Class A Fires

Fires involving ordinary combustible material, such as
- wood
- paper
- cloth
- rubber

Class B Fires

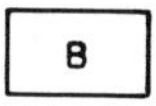

Fires involving flammable liquids and gases, such as
- petroleum products
- gasoline
- oil
- grease

Class C Fires

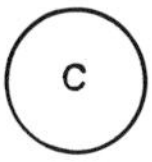

Fires involving electrical equipment, such as
- transformers
- wires
- fuse box
- switches
- control panels
- appliances

Class D Fires

Fires involving combustible materials, such as
- magnesium
- titanium
- metallic sodium

Note: Some metals combustible under certain conditions are
- calcium
- zinc
- aluminum

Figure 4–1 Classification of fire extinguishers and fires.

The fire guard should then report the following:
the exact location of the fire/smoke condition
the severity of the fire/smoke condition
whether or not he has begun the evacuation process
whether or not the fire/smoke condition is extinguishable or containable
any special problems encountered

Table 4–1   Fireguard inspection checklist

---

1. Inspect and ensure that all exits, stairways and hallways are free of obstruction and available for use.
2. Examine all doors to determine if they are operative and available for use.
3. Inspect and ensure that all self-closing doors are unobstructed and kept closed.
4. Continually inspect the facility for accumulation of rubbish and debris.
5. Know the location of all fire extinguishers and know how to operate them.
6. Determine and report all noticeable fire or safety hazards.
7. Ensure that exits, stairwells, and floors are properly identified by a number or letter designation.
8. Ascertain that reentry from stairways is provided for on each floor.
9. Check all emergency lighting to be sure it is functional.

Depending on the severity of the fire/smoke condition, the fireguard should then proceed to the fire floor and the floor *above* the fire to alert the occupants of the present condition. He should then attempt to evacuate the occupants of the fire floor via the safest route—usually a stairway.

It is imperative that fireguards be familiar with the interior and exterior of the facility. They should know where all the manual pull stations are located as well as the stairwells and the sprinkler shut-off valve. During their routine patrol of the interior of the facility, it is their primary function to observe and report fire conditions or hazards. Their inspection of the facility should be monitored on a daily basis. Table 4–1 is an example of a fire guard's inspection checklist that should be reviewed by the FSD or the DFSD daily. Depending on your facility, you may add to or subtract from the inspection list to accommodate the variables of your facility.

Fire Fighter

In the event of a fire at the facility, the primary function of the fire fighter, upon hearing the fire alarm, is to respond to the floor on which the fire condition has been reported by using the stairs. Upon reaching the floor, the fire fighter should determine whether that floor may be entered safely. If your organization has a large staff, the fire figther should be accompanied by the evacuation officer.

Prior to entering the fire floor from the stairwell, the fire fighter should determine if the amount of smoke, fire and/or heat presents a threat to his life or others. The fire figther should then notify the FSD at the fire control station of the following information using her two-way radio.

Notify the FSD upon reaching the fire floor.

Describe the conditions on that floor, such as the extent of the fire and the severity of the fire, heat, or smoke.

Inform the FSD if occupants are on the fire floor.

Inform the FSD if these occupants can be evacuated safely.

Inform the FSD if the fire is extinguishable and/or containable.

The fire fighter must determine whether or not the fire can be approached safely. If it can, efforts should be made to contain, isolate, and extinguish the fire. The fire fighter must immediately report her determination to the FSD at the fire control station.

A secondary consideration of the fire fighter is to isolate the fire as much as possible by ensuring that all doors and windows on the fire floor remain closed. Fire figthers should also assist the evacuation officer in evacuating all people from the fire floor. If this cannot be safely accomplished, occupants should be instructed to remain at their present locations with the doors closed. However, many times people do not want to remain at the fire location because of the X-factor discussed in Chapter One.

It is imperative that the FSD be advised if persons are trapped on the fire floor. It is the FSD's responsibility to immediately inform the fire department of this fact upon their arrival.

Fire fighters should remain either on the fire floor, or in the stairwell of the fire floor, until the arrival of the fire department. They must provide the fire department with all information regarding the fire condition, such as the location and severity of the fire, and whether or not inhabitants are trapped on the fire floor. When the fire department arrives at the fire floor, the fire fighter must notify the fire control station. The fire fighter should remain either on or in the vicinity of the fire floor until told to vacate the area by the fire officer, FSD, or DFSD. When directed, the fire fighter should return to the lobby of the facility to assist in crowd control.

Evacuation Officer

Upon hearing a fire alarm, the evacuation officer must proceed immediately to the fire floor and assist the fireguards and fire fighters in evacuating all personnel from the fire floor as well as the floor *above* the fire. Prior to entering the fire floor, the evacuation officer should determine whether the floor can be entered safely. The presence of a heavy fire or a heavy smoke condition, which would endanger the life of the officer, is the determining factor as to whether or not he should enter the floor.

If the evacuation officer determines it is safe to enter the fire floor, he should direct occupants to the fire stairs. The evacuation effort must be coordinated with the fireguard and FSD to determine which fire stairs are safe to use. The evacuation officer should ensure that all doors are closed after individuals exit their locations and that no individuals use the elevators, with the exception of the fire department.

When total evacuation of the fire floor is complete, the evacuation officer must inform the FSD at the fire command station. The evacuation officer should then proceed to the main lobby of the facility and assist in crowd control and maintaining order.

Direction Officer

When fire department fire fighters arrive at the facility anticipating a blazing fire, their primary concerns are to prevent any loss of life or injury and to minimize property loss. Since they are not as familiar with the facility as are members of your security/fire safety department, it is important for your personnel to relay precise information to fire department personnel on their arrival.

Using the elevator banks and the fire stairs as points of reference, the location of the fire should be described. Any other information relative to the fire should be given to the fire department, such as the nature and extent of the fire, whether attempts have been made to extinguish or contain the fire, the success of these attempts, and whether evacuation is complete or inhabitants remain trapped on the fire floors.

The direction officer should have additional copies of the facility's floor plans ready to give to the fire department. Master keys should be provided to the direction officer in the event that the FSD cannot get to the fire department upon their arrival. At the direction of the fire officer, the FSD, or the DFSD, the direction officer should assist other members of the fire brigade in crowd control and maintain order.

Communication Officer

The communication officer has the most significant and important function in the fire operation plan. This officer is solely responsible for maintaining communication with the members of the fire brigade before, during, and after a fire or smoke condition. This officer must transmit, receive, and record information relative to the fire or smoke condition. Accurate recording of information is crucial to the FSD and the fire department, and is especially useful when it is time to write the incident report. The recording of times, names, ladder company numbers, and engine company numbers is vital to the accuracy of the FSD's overall report.

Upon receiving a report of a fire or smoke condition, the communication officer should ensure that

the FSD and DFSD have been advised of the fire or smoke condition

members of the fire brigade have been dispatched to their respective locations

members of the fire brigade are in immediate communication with the fire command station upon their arrival at the fire or smoke floor

The communication officer should issue updates to the FSD every 5 minutes regarding the fire/smoke condition. During a fire/smoke condition, all other functions normally performed by the dispatcher, who now assumes the responsibility of the communication officer, become of secondary importance unless, or course, those functions deal with another life threatening situation also in existence.

COMMUNICATION EQUIPMENT

The communication equipment owned and used by your facility depends solely on what you inherit or buy. Remember, city ordinances are mandated and compliance is required. Therefore, if the city requires you to have certain equipment, it is in the best interests of the organization to comply with the fire or building codes.

In general, most security/fire prevention departments have a control or *nerve center* from which guards communicate with one another and local emergency service departments via two-way radio or telephone. If your department does not have a communication center, include one in your budget and support your proposal with documentation and appropriate justification.

Assume you do not have a communication system and your nerve center must be manned 24 hours a day. The facility must conform to the local fire and building codes.

> Where the equipment for guard communication, including guards on watch patrol, requires that signals from guards be monitored, the control center should be provided with an operator. Additional operators and twenty-four hour operator service should be provided at the control center according to the character of the guard service provided. For some services, runners or guards who can be dispatched to investigate signals should also be provided.[17]

In order for a communication system to be used effectively, adequately trained personnel must be your first priority. The training implemented by the FSD is based on the sophistication of the system, the quality of the staff, and the procedures developed by the FSD.

The public address system used by the facility shall be capable of being heard easily (customarily, 80 decibels at 60 feet) in all existing corridors, hallways, passageways, and stairs.

Most fire codes require that your facility have an adequate and reliable fire communication system. The communication system is a pivotal link between life and death during a fire. Your facility should be equipped to communicate via two-way radio or telephone with your fire command station. In various cities, some fire codes for hotels require a permanent telephone at the fire command station and telephones or telephone jacks on other floors near the main riser. These telephones or telephone jacks must be secured in a safeguarded area that can be opened with a fire department standard key. Fire communication systems must meet with the approval of Underwriters Laboratories and the NFPA. "A fire communication slotted coaxial cable radio system is installed to provide adequate communication capability throughout a building. Adequate communication is defined as the capability for clear two-way communication between a fire department portable radio at the lobby command post and another fire department radio at any other point in the building. Such a system shall be acceptable to the fire department."[18]

Every emergency communication system should be run on a dependable, uninterrupted, charged energy supply. This energy supply should be resilient enough to continue operating during a fire or smoke condition. As the FSD planning your communication system, what foreseeable problems can your communication system cause you? "The problems associated with the installation and protection of the communication equipment are: (1) that the equipment often introduces combustible fuel loading into an otherwise fire safe area of the building, and (2) that the installation of the equipment can lead to serious reduction of the structural compartmentation originally designed into the facility."[19]

The communication system your facility used must meet with the specific requirements established by the fire and building codes. The terms *detection, alarm,* and *communication* (DAC) are part of every security FSD's vocabulary. If your security personnel are adequately trained in fire safety, they will know that as soon as fire or smoke is detected, they should activate an alarm. The information should be communicated in a clear, precise fashion so that your fire operation plan can go into effect. Time is the critical factor in fire control and containment. A potential deadly fire can be controlled or contained at the outset if the DAC concept as outlined above is followed properly. The next course of action solely depends on your building classification and your fire operation plan.

Communications are vital to fire control and containment. Communications notify occupants and employees, render sufficient directions for action, and forewarn the fire department of the fire condition. Too often, fire department notification has been postponed because occupants were under the false conception that a local alarm notified the fire department.

SUMMARY

Formulating, designing, and implementing the duties of the fire brigade is a difficult and time-consuming responsiblity. The type of your facility as well as its classification must be taken into consideration when determining the duties of the fire brigade.

The FSD and DFSD are ultimately responsible for choosing and training qualified candidates for the fire brigade. Many times, selecting qualified fire brigade candidates becomes an enormous asignment. After you design the duties of the fire brigade, it is your responsibility to ensure that they carry out their instructions. This is achieved through training, preparation, and education. The fire drills conducted by the FSD will help to measure the performance of the fire brigade.

The members of the fire brigade (i.e., alarm box runner, backup runner, fireguard, fire fighter, evacuation officer, direction officer, and communication officer) are all vital components of a team. If these individuals are properly trained and educated in all fire brigade responsibilities, you have the makings of a sound and effective fire safety department.

5

Fire Detection and Alarm Systems

As the security/FSD, have you ever really paid much attention to the architectural layout of your facility? You are probably familiar with security alarms, closed-circuit televisions, and electric article surveillance, but have you ever considered what it would take and how to begin implementing and using a fire detection and alarm system?

Prior to taking on a task of this size by yourself, put your investigative instincts to work and research fire detection and alarm systems and determine the type of system you need for your facility. Make inquiries at your local fire department. Next, contact a few reputable alarm installation and service companies. Do not be shy when asking about these systems and their effectiveness.

If your facility is without a fire detection and alarm system, and you need to have one installed, you first need to know the building's classification. When you have this information, ask fire detection and alarm system representatives what types of systems their companies sell and determine whether any of the systems are sufficient for your building's classification.

Lawrence G. Visotsky, sales representative of Simplex in New York City, sees alarm companies growing by leaps and bounds over the next 5 years, due to the legislation upgrading fire laws and building codes. He states, "Today many companies are not sure if the fire or building laws apply to them. When they are informed that they do, their basic approach is how much will this system cost? Unfortunately, it is not like buying a car with various features. The fire alarm systems must detect, alarm, and communicate (DAC) from the point of origin to a central station."[20]

Some fire detection systems are better than others. However, you do not want a system that is so sophisticated that your staff will have problems responding to the signals being sent. The approved system you select for your facility should be able to automatically transmit a fire alarm or smoke condition signal to your central station, an alarm company's central station, and/or

45

directly to the local fire department. The system you choose depends upon your operation plan and your budget.

Your fire system should be a distinct, electronically supervised, sanctioned system that includes an automated interior alarm and voice communicating system so that any interior alarm that sounds will identify its location at the fire command station, the mechanical master midpoint, and the regularly assigned location of the FSD.

The hardware and wiring of the system shall meet the approval of the municipality's fire commissioner. Again, the importance of choosing a reputable company to install approved and accepted fire systems is clearly evident.

Most fire systems use audible and visual signals when transmitting an alarm. Your fire system should be capable of silencing these signals from your fire command station. The fire command station should allow the floor stations to generate information over the communication system. Your system should work in conjunction with your communication system and be capable of making announcements during a fire condition.

TYPES OF SYSTEMS

There are various fire detection and alarm systems in use throughout the world today. One company will say their system is the best, while another will say their system is the least expensive. You must be sold on a system that achieves its overall purpose without depleting your budget.

The NFPA lists standards and codes with which alarm manufacturers must comply. If your system is not in compliance with NFPA codes, you will be installing a nonapproved fire system.

Fire alarm signaling systems are issued classifications in accordance with their primary functions. The NFPA requires that a system's installation, maintenance, and use conform to the following basic criteria established for fire detection and alarm systems.

1. A control unit.
2. A primary (main) power supply which usually is the local light and power service.
3. A secondary (standby) power supply.
4. One or more initiating device circuits to which automatic fire detectors, manual fire alarm boxes, water flow alarm devices, and other alarm initiating devices are connected.
5. One or more alarm indicating device circuits to which alarm indicating signals such as bells, horns, speakers, etc., are connected, or to which an off premises alarm is connected, or both.[21]

Local System

The term *local system* means what it implies. If your fire system is activated, it will send an alarm throughout your facility, but not to the fire department or an alarm-monitoring company. The primary purpose of a local system is to warn occupants in your facility that an evacuation procedure may be necessary. If your facility is unoccupied after certain hours and the alarm registers indicating a fire or smoke condition, chances are the fire or smoke will cause much damage due to nonnotification of the fire department.

The power supply of your local system should operate the system for 24 hours under an average load and then maintain the alarm system for an additional 5 minutes. This criterion covers a regular day of occupancy and a 5-minute alarm to evacuate the building.

Auxiliary System

According to Charles E. Zimmerman, PE, "an auxiliary protective signaling system has circuitry connecting the alarm initiating apparatus to the municipal fire alarm system, either through a nearby master fire alarm box or through a dedicated telephone line run directly to the municipal communication switchboard. The signal received by the local fire department is the same received when an individual manually activates the fire alarm box. Because the fire department recognizes which municipal boxes are auxiliarized, responding fire companies can check for an alarm originating within the facility."[22]

Remote Station System

Justifications within your budget sometimes afford you the opportunity to add services to your fire operation plan, such as adding a remote station system. The remote station system is similar in function to an auxiliary system, but different in that the alarm transmitted from this system is received at a remote location attended by trained and supervised personnel 24 hours a day.

When the alarm is triggered, remote station personnel act as the intermediaries and notify the fire department via telephone. The signal is transmitted over a dedicated telephone line and registers audibly and visually at the remote station. The system also signals automatically at the facility's fire command station, thereby alerting in-house personnel of a fire or smoke condition.

If the system is transmitting false or incorrect signals, the remote station will contact you to notify you of a potential problem.

Underwriters Laboratories requires that the remote station have an independent secondary power supply that operates the system up to 60 hours, followed by 5 minutes of alarm.

Proprietary System

The proprietary system has a designated fire command station manned with personnel, a receiving console, and transmitting units. The receiving console receives signals from within the facility when a fire or smoke condition registers. Due to advanced technology and the use of computers and electronics, it is wise to ask a distributor if signal multiplexing is required for your type of facility and your type of system.

Proprietary receiving consoles can vary in size, shape, color, and cost. The system you select should consist of specific lights, a digital display, or a CRT optical exhibit that indicates the fire's point of origin, signals with an audible alarm, and provides a hard copy printout.

Your system should be able to perform specific tasks during a fire. If the system is equipped to control smoke in the facility by automatically opening and closing dampers in heating, ventilation, and air condition (HVAC) systems, and turning on exhaust fans, your system is most likely tied into the facility's computer system. This means that the building's engineers are more likely to act as members of the fire brigade than members of your staff.

Some systems are so advanced electronically that the facility's elevators are also integrated with the fire control system. Some systems can adjust elevator controls so that, in the event of a fire or smoke condition, the elevators bypass the fire floors and automatically return to the lobby.

Central Station System

Proprietary and central station systems are similar in design and concept. However, the central station receiving the alarm signal has no direct knowledge of the facility that is sending the signal, while the proprietary station has employees or agency security officers manning the fire command station.

Mr. Visotsky of Simplex, Inc. further states, "the central station has trained personnel to monitor and man the central station twenty-four hours a day. Upon receiving a signal from a client's facility, the operator at first tries to establish contact with the client. The central station will advise the client that an active alarm has been transmitted and asks if the client wishes the local fire department notified."[23] If the central station operator has instructions from the client to immediately notify the fire department, the operator will follow those instructions.

The main difference between the central station and the proprietary system is the signal transmission between the facility and the receiving station. The most widely known circuit in a central receiving station is the McCulloh circuit.

> The McCulloh circuit, the oldest transmitting means in central station use, normally transmits over two wires but can be switched manually or automatically to transmit over one wire and ground. With this capability, the system will not be rendered inoperative by a break or ground fault in a single wire. It is customary to connect the plants of several subscribers to a single transmission circuit. Each such circuit transmits in a recording instrument, and each subscriber has one more coded signal number not repeated on that particular circuit.[24]

The client is also known as the subscriber and is charged a monthly fee by the central station monitoring the client's facility. The subscriber must understand that problems in the system of excessive false alarms can cause the central station to discontinue or penalize your facility.

The operators at the central station are usually issued identification numbers. When speaking with them, be sure you get their identifiction number. It is important to establish a rapport with the operators at the central station. Remember, they are following your fire operation plan and must be advised by you of the procedures you want followed.

A typical fire command station with a console will have the following components: (See Figure 5–1)

- alarm/trouble zone lamps
- speaker zone select switches
- speaker zone select lamps
- telephone zone select lamps
- telephone select switches
- control panel lamps
- master control panel switches

Any additional components used throughout your facility depends on your system or fire code mandates. Additional components can include

- hard-wired ionization smoke detectors
- hard-wired heat detectors
- battery-operated smoke detectors
- battery-operated heat detectors
- manual alarm pull stations
- remote station telephone jacks
- alarm/communication speakers
- tamper alarms

Figure 5–1 Fire communication controller/command station. (Edwards 5800, Edwards Co.)

- water flow alarms
- automatic smoke doors
- auxiliary annunciator panel

During fire drills or when inspecting your system, be sure that system testing is in accordance with the criteria established by your local fire authority. If your city has no conflicting local regulations regarding the testing of your fire system, it is best to conduct monthly tests on your fire system.

Most fire drills can be initiated from any manual pull station throughout your facility. Most systems have an alarm test feature in the system. Always notify your remote and central stations when you are conducting fire drills.

NATIONAL FIRE PROTECTION ASSOCIATION

The NFPA was established in 1896 by insurance companies and a leading manufacturer of automatic sprinklers. The original goal of the organization was to develop standards for sprinkler use and installation. After these standards were established, other regulations followed to regulate fire hoses, fire doors, and hydrants.

The NFPA and Underwriters Laboratories use consensus procedures to institute and develop ethics. The NFPA does not actually test all fire apparatus.

The standards established by the NFPA cover various categories in the fire safety field. You need not know all the standards, however it is imperative that you know where to find information pertaining to a specific standard. *The Fire Protection Handbook* is the FSD's bible. The first edition was printed in 1896; in 1986, the sixteenth edition was printed. This book is used to educate and inform all individuals in the security and fire safety fields.

Life Safety Code

The *Life Safety Code* was adopted by the NFPA in November 1980. In 1942, in Boston, Massachusetts, the Coconut Grove Night Club Fire claimed 492 lives. This fire focused national attention on the importance of an adequate number of exits and related fire safety features. Due to this fire, the *Building Exits Code* was established. In 1966, the title was changed to *Code for Safety to Life from Fire in Buildings and Structures*. Today, the code is most commonly referred to as the *Life Safety Code*.

The foremost objective of the Life Safety Code is establishing minimum requirements that afford a reasonable degree of safety from fire in buildings and other structures. As the FSD, it is important to remember that each building is classified differently, usually based on relative degree of fire and life safety risk. The Life Safety Code applies to your facility. Each facility designed for human occupany *must* have ample number of exits to allow immediate evacuation of the building's occupants in the event of a fire or catastrophe.

The code specifies that all exits shall be arranged and maintained to provide egress from all parts of the building when occupied. No locks or fastening devices shall be installed to prevent escape from the interior of any facility. As the security/FSD, this presents a burden to you from a security standpoint. How can you remedy this situation? Electromagnetic locks are new concepts to legally lock fire exit doors and various companies distribute these locks. As the FSD, you require a fail-safe locking system for fire exit doors that tie into your fire detection system.

One system on the market today, which conforms to specific codes, is the *Andralock System*. (See Figure 5–2) This system is presently used in hospitals, commercial buildings, schools, and museums. The Andralock System is approved by the New York City Board of Standards and Appeals, the New York City Fire Department, and the New Jersey Department of Education. It is also New Jersey Uniform Fire Safety Code-qualified and Underwriters Laboratory-listed. This system meets the NFPA Life Safety Code 101, as amended in 1981, and also meets the BOCA code.

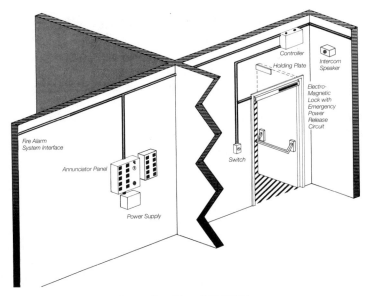

Andra Systems Inc. 74 Warren Street, Brooklyn, NY 10007.

Figure 5–2   The Andralock Exit-door system.

Each Andralock System is customized for a particular environment. The system is designed to meet NFPA standards and local fire department codes and regulations. Normally, fire exit doors are equipped with panic bars that allow easy access by intruders. This system overrides the panic hardware, but when the power is removed the lock instantly releases and the panic hardware is once again the only securing force. When this system is incorporated in a facility's fire system, it automatically releases in the event of a fire, thus presenting a fail-safe option for use on fire exit doors.

The Life Safety Code also mandates that exits and routes to exits be clearly marked. If artificial illumination or emergency lighting is required in your facility, be sure they are operational and located where specified.

Interior fire alarms must be provided, where necessary, to alert occupants in the event of a fire. Every vertical exit or aperture between floors of a facility must be properly enclosed or protected to ensure reasonable safety to inhabitants using exits and to avert the spread of fire, smoke, or fumes through vertical openings.

At a minimum, two means of egress shall be provided in every facility in which the facility's dimensions or occupancy jeopardizes inhabitants attempting to use a single means of egress obstructed by fire, smoke, or fumes. These two means of egress are required to lessen the possibility that all exits are impassable during an emergency condition.

The requirements of the Life Safety Code are important and compliance is a must. The code was established to provide *minimum* requirements that provide a reasonable degree of safety from fire in buildings and other structures.

SUMMARY

The proper installation and use of fire safety and communication systems are vital tasks for the FSD. Actual fire system installation will be done by a reputable alarm company. The functions of the fire system must be thoroughly reviewed by the FSD. The system must meet the requirements of the NFPA as well as the Underwriters Laboratories. If the system is too sophisticated or inadequate, it is useless to your organization. If you inherit a fire safety system, be sure it meets the standards and codes of the NFPA, with which alarm manufacturers must comply.

NFPA standards are crucial to ensuring fire safety. NFPA recommendations regarding fire safety and fire standards are reviewed and tested daily throughout various cities.

As the security/FSD, the facility for which you are responsible for protecting is, in one way or another, covered by the NFPA Life Safety Code. The NFPA has formed subcommittees based on various building types. The subcommittees recommend and review life safety codes in buildings and other structures.

Other organizations that impact the fire safety and life safety of your building are BOCA, ICBO, and SBCCI. The codes established by these organizations are for upgrading a facility as well as creating uniform standards for fire safety.

You must be familiar with egress locking systems that can be used in conjunction with your fire system. Your fire system should conform to NFPA standards and local fire department codes.

As you become familiar with your city's fire and building codes, you can begin to implement your fire operation plan. You will spend a lot of time researching codes and ordinances. Seek the advice of qualified persons in the fire safety field prior to making any major equipment purchases. Seek advice if you are unsure of some codes.

6

Conflicting Codes and How to Deal with Them

The majority of the building codes throughout the United States are based on model codes or other nationally acknowledged standards. Building departments offer code books or pamphlets for sale or code books can be obtained from the organizations that publish the standards.

You may be overwhelmed at the thought of knowing and memorizing each and every code. There is no need to memorize these codes, but it is important to know about them and to know to which source to refer in the event of a problem. Remember, you will be interfacing with your facility's building operations department when establishing your fire operation plan. This department is responsible for complying with the majority of building codes with which you are unfamiliar. It is crucial to know the various codes and know who is responsible for implementing them.

THE FOUR MODEL BUILDING CODE GROUPS

There are four building code groups that establish model codes. You should become familiar with them.

1. American Insurance Association (AIA)
2. Building Officials and Code Administrators
3. International Conference of Building Officials
4. Southern Building Code Congress International

Each of these organizations has developed or designed model building codes. There are other code-related organizations that we will discuss, but these four have had a tremendous impact on the security and fire safety industries.

55

American Insurance Association (AIA)

The AIA was originally known as the National Bureau of Fire Underwriters and published the first national building code in 1905. The code was designed and used as an archetype to be adopted by cities, and was a reason to study building regulations of towns and cities for rating purposes.

In 1976, the AIA completed the last code revision in that year's edition and announced that the National Building Code would no longer be updated. AIA also indicated that BOCA had acquired the right to use the term *National Building Code.*

Building Officials and Code Administrators

BOCA was responsible for publishing the first edition of the *Basic/Building Code* (BBC) in 1950. This organization also publishes plumbing, mechanical, and fire prevention codes. These codes are subject to review and revised annually if needed.

> The Basic/National Building Code was initially prepared and has been maintained on the premise that all matters pertaining to the construction of a building and built into it, either in its initial construction or through subsequent alterations, repair or extension, should be covered by the building code. This includes fire protection of the building elements as well as fire separation walls or other precautions required for protection against specific hazards of the particular use of the building.[25]

The purpose of this code is to adequately provide the essential responsibilities of the fire official and building official, delineate their accountabilities and liabilities, and require that they work together to provide reasonable safety in all buildings under their jurisdiction.

BOCA national building codes are designed to be referenced by state and local municipalities only. Jurisdictions adopting these codes may make essential additions, deletions, and amendments in the legislation they choose to enact.

International Conference of Building Officials

The ICBO published the first uniform building code in 1927, which is primarily used in the western United States. ICBO also publishes mechanical and plumbing codes, and a fire prevention code in association with the Western Fire Chiefs Association. Code changes are made annually and an amended version of the codes is published every 3 years.

The Conference's stated objectives relate to the development and publishing of regulations and educational materials designed to increase the standardization of building construction regulations and the enforcement of these regulations. To encourage this standardization, the Conference maintains a staff of architects and engineers in Whittier, CA, which provides governmental bodies with technical assistance in the administration and enforcement of the ICBO codes.[26]

Southern Building Code Congress International

The SBCCI was established in 1940 and published the first southern standard building code in 1945. Today, this code is known as the *Standard Building Code* (SBC). The SBC is primarily used throughout the southern portion of the United States.

The SBCCI is similar to the ICBO and BOCA in that it publishes mechanical, plumbing, fire prevention, and gas codes. The codes are modified and reprinted every 3 years. The primary purpose of the SBCCI is to elevate standardization in construction regulations and enforcement. This organization has its technical offices in Birmingham, Alabama.

You will no doubt be faced with the situation of having to know when one code is superseded by another. You must remember that the association of building code requirements is customary for *new* construction or for significant alterations to existing structures. For example, if you received a building permit in 1988 but started actual construction in 1990, provisions must be made to upgrade your facility to the 1990 standards. Retroactive application of code requirements is infrequent. Building code applicability typically terminates with the issuance of an occupancy permit.

> The basic premise that fire legislation should regulate for the safety of current occupants and for current risk is not generally the province of building codes after occupancy. After occupancy fire codes, or more precisely, fire prevention codes, apply. It is also at the point that the authority of the building official usually ends and that of the fire official begins.[27]

Most municipalities seek interaction between the fire department and building department during construction and after occupancy of a building. The way a particular city decides to have the two agencies interact is entirely up to that city. Many *states* also differ in their legislation of building and fire codes. Some states may have local municipalities that are solely responsible for code enforcement, while in other states building codes are mandatory. Still other states establish the minimum state code for a municipality. Be familiar with your state's and city's regulations regarding code enforcement.

DEALING WITH VARIOUS CODES

In most instances, your facility will have building engineers, property managers, or a facility manager who are responsible for dealing with the fire or building departments. Interpretation of the various codes should be discussed, in detail, with the FSD and the individual responsible for the entire building.

According to William F. Collins, an AIA architect and principal of his own company located in Setauket, New York, there are many codes throughout the United States. There are also different codes in each state. An excellent example would be New York State. The majority of the state uses the New York State Uniform Fire Prevention and Building Code. However, the City of New York formulated their own codes in The Building Code of the City of New York. Then again, some local villages have adopted "BOCA" as their building code.[28]

Collins goes on to state that although there are differences within fire codes, they all try to accomplish the same goal—protecting the safety, health, and welfare of the general public. Collins makes an excellent point in stating that security professionals who become FSDs should keep one important factor in mind—fire codes and building codes are to be used in conjunction with one another. These codes work together to protect life and property. As the FSD, you must be cognizant of both fire and building codes.

The 1989 Directory of Building Codes and Regulations, Volume IV State Commercial Codes, which is published by the National Conference of States on Building Codes and Standards, Inc., details the building, fire, and life safety codes for the 50 states, District of Columbia, Puerto Rico, and the Virgin Islands, some of which are outlined on the pages remaining in this chapter. This book also lists mechanical, plumbing, electrical, accessibility, and energy codes, amendments, and enforcement of the codes. The following list contains some of the abbreviations used in the Directory of Building Codes and Regulations.

ANSI A117.1—Making Buildings and Facilities Accessible to and Usable by Physically Handicapped People

NBC—National Building Code (or Basic Building Code)

NEC—National Electrical Code

NFC—National Fire Prevention Code (or Basic Fire Prevention Code)

UBC—Uniform Building Code

UFC—Uniform Fire Code

UHC—Uniform Housing Code

NFPA 101—Life Safety Code

NFPA 501A—Firesafety Criteria for Manufactured Home Installations, Sites, and Communities

SBC—Standard Building Code

SFC—Standard Fire Prevention Code

SHC—Standard Housing Code

STATE CODES AND REGULATIONS

Alabama

Building Code: 1988 SBC

Enforcement of Building Codes: Attorney General of the Director of the State Building Commission, 800 S. McDonough Street, Montgomery, AL 36130; 205–261–4082

Fire Prevention Codes: 1985 SFC and 1985 NFPA codes

Enforcement of Fire Codes: State Fire Marshal, 135 S. Union Street, Room 140, Montgomery, AL 36130; 205–269–3575

Life Safety Code: 1985 NFPA 101

Enforcement of Life Safety Codes: same as fire codes

Alaska

Building Code: 1985 UBC

Enforcement of Building Codes: State Fire Marshal, Alaska Fire Prevention Division, 5700 E. Tudor Road, Anchorage, AK 99507–1225; 907–269–5604

Fire Prevention Code: 1985 UFC

Enforcement of Fire Codes: same as building codes

Life Safety Code: none

Enforcement of Life Safety Codes: none

Arizona

Building Code: none

Enforcement of Building Codes: none

Fire Prevention Code: 1985 UFC

Enforcement of Fire Codes: none

Life Safety Code: none

Enforcement of Life Safety Codes: none

Arkansas

Building Code: 1982 SBC

Enforcement of Building Codes: Commander of State Police, Fire Marshal Section, #3 Natural Resource Drive, Little Rock, AR 72215; 501–224–3103

Fire Prevention Code: 1982 SFC

Enforcement of Fire Codes: same as building codes

Life Safety Code: 1981 NFPA 101 by reference
Enforcement of Life Safety Codes: same as building codes

California

Building Code: 1988 UBC (effective 7/89 state, 1/90 local)
Enforcement of Building Codes:
  For Clinics and Health Facilities: Office of Statewide Health Planning & Development, 1600 9th Street, Room 410, Sacramento, CA 95814; 914–322–3212
  For Education and Institutional Buildings: Deputy State Fire Marshal, 7171 Bowling Drive, Suite 600, Sacramento, CA 95823; 916–427–4166
  For Public Schools: Office of State Architect, 400 P Street, Fifth Floor, Sacramento, CA 95814; 916–445–5939
Fire Prevention Codes: 1988 UBC, reference 1988 UFC and NFPA
Enforcement of Fire Codes: Deputy State Fire Marshal, 7171 Bowling Drive, Suite 600, Sacramento, CA 95823; 916–427–4166
Life Safety Code: none
Enforcement of Life Safety Codes: none
  Note: The fire marshal enforces codes for hospitals and nursing facilities under a federal contract for the Department of Health. The fire marshal also enforces NFPA 101 codes for hospitals and nursing facilities under a federal contract with the Department of Health.

Colorado

Building Code: none statewide, promulgated locally
Enforcement of Building Codes: local
Fire Prevention Code: none
Enforcement of Fire Codes: none
Life Safety Code: none
Enforcement of Life Safety Codes: none

Connecticut

Building Code: 1984 NBC, with 1986 supplements
Enforcement of Building Codes: local
Fire Prevention Code: same as life safety codes
Enforcement of Fire Codes: same as life safety codes
Life Safety Code: 1985 NFPA 101
Enforcement of Life Safety Codes: Deputy State Fire Marshal, Division of State Police, 294 Colony Street, Meriden, CT 06450; 203–238–6620

Delaware

Building Code: none
Enforcement of Building Codes: none
Fire Prevention Code: 1988 NFPA 1
Enforcement of Fire Codes: State Fire Marshal's Office, RD 2, Box 166A, Dover, DE 19901; 302–736–4393
Life Safety Code: 1985 NFPA 101
Enforcement of Life Safety Codes: same as fire codes

Florida

Building Code: 1988 SBC, 1988 South Florida Building Code, EPCOT Code
Enforcement of Building Codes: local
Fire Prevention Code: 1985 Standard Fire Prevention Code
Enforcement of Fire Codes: local
Life Safety Code: 1985 NFPA 1
Enforcement of Life Safety Codes: local

Georgia

Building Code: 1988 SBC
Enforcement of Building Codes: local
Fire Prevention Code: state fire marshal regulations based on NFPA
Enforcement of Fire Codes: local or state, Fire Marshal, 2 Martin Luther King Jr. Drive SW, West Tower, Suite 620, Atlanta, GA 30334; 404–656–9498
Life Safety Code: 1985 NFPA 101
Enforcement of Life Safety Codes: same as fire codes

Hawaii

Building Code: State Housing Code Chapter 14 Title 11, Administrative Rules
Enforcement of Building Codes: local
Fire Prevention Code: none statewide, promulgated locally
Enforcement of Fire Codes: local
Life Safety Code: none statewide, promulgated locally
Enforcement of Life Safety Codes: local

Idaho

Building Code: none
Enforcement of Building Codes: none

Fire Prevention Code: 1985 UFC
Enforcement of Fire Codes: State Fire Marshal, Department of Insurance, 500 S. 10th, Boise, ID 83720; 208–334–3808
Life Safety Code: none
Enforcement of Life Safety Codes: none

Illinois

Building Code: none
Enforcement of Building Codes: none
Fire Prevention Code: same as life safety code
Enforcement of Fire Codes: same as life safety codes
Life Safety Code: 1985 NFPA 101
Enforcement of Life Safety Codes: Acting Chief, Division of Fire Prevention, Office of State Fire Marshal, 3150 Executive Park Drive, Springfield, IL 62703; 217–786–7180

Indiana

Building Code: 1988 UBC
Enforcement of Building Codes: Acting State Building Commissioner, Department of Fire & Building Services, 1099 N. Meridian Street, Suite 900, Indianapolis, IN 46204; 317–232–1400
Fire Prevention Code: 1988 UFC
Enforcement of Fire Codes: State Fire Marshal, Department of Fire & Building Services, 1099 N. Meridian Street, Suite 900, Indianapolis, IN 46204; 317–232–2226
Life Safety Code: none
Enforcement of Life Safety Codes: none

Iowa

Building Code: 1988 UBC
Enforcement of Building Codes: State Building Code Bureau/Department of Public Safety, Wallace State Office Building, Des Moines, IA 50319; 515–281–3807
Fire Prevention Code: Iowa fire marshal rules and regulations, reference NFPA standards
Enforcement of Fire Codes: State Fire Marshal, State Building Code Bureau/Department of Public Safety, Wallace State Office Building, Des Moines, IA 50319; 515–281–5821
Life Safety Code: same as fire prevention code
Enforcement of Life Safety Codes: same as fire prevention codes

Kansas

Building Code: none
Enforcement of Building Codes: none
Fire Prevention Code: none
Enforcement of Fire Codes: none
Life Safety Code: 1988 NFPA 101
Enforcement of Life Safety Codes: State Fire Marshal, 700 SW Jackson Street, Suite 600, Topeka, KS 66612; 913–296–3401

Kentucky

Building Code: 1987 NBC
Enforcement of Building Codes: local or Director, Division of Building Code Enforcement, The 127 Building, US 127 South, Frankfort, KY 40601; 502–564–8090
Fire Prevention Code: 1988 Fire Safety Standards for Existing Buildings, reference in the Building Code
Enforcement of Fire Codes: State Fire Marshal, The 127 Building, US 127 South, Frankfort, KY 40601; 502–564–8044
Life Safety Code: 1985 NFPA 101—Health & Day Care, State Building Codes—New Commercial Buildings, 1976 NFPA 101—Existing Buildings
Enforcement of Life Safety Codes: State Fire Marshal, Department of Housing, Buildings and Construction, The 127 Building, US 127 South, Frankfort, KY 40610; 502–564–3626.

Louisiana

Building Code: none
Enforcement of Building Codes: none
Fire Prevention Code: 1988 NFPA 101, all NFPA standards references in Chapter 32
Enforcement of Fire Codes: local or Chief Architect, 7701 Independence Boulevard, Baton Rouge, LA 70806; 504–925–4920
Life Safety Code: same as fire prevention code
Enforcement of Life Safety Codes: Chief Architect, 7701 Independence Boulevard, Baton Rouge, LA 71806; 504–925–4920

Maine

Building Code: none
Enforcement of Building Codes: none
Fire Prevention Code: none
Enforcement of Fire Codes: none
Life Safety Code: 1988 NFPA 101
Enforcement of Life Safety Codes: State Fire Marshal, State House Station #52, 317 State Street, Augusta, ME 04333; 207–289–2481

Maryland

Building Code: none
Enforcement of Building Codes: none
Fire Prevention Code: 1984 NFC
Enforcement of Fire Codes: Chief Fire Protection Engineer, 6776 Reisterstown Road, Suite 314, Baltimore, MD 21215; 301–764–4324
Life Safety Code: 1988 NFPA 101
Enforcement of Life Safety Codes: same as fire codes

Massachusetts

Building Code: 1988 State Building Code
Enforcement of Building Codes: local
Fire Prevention Code: 1978 NFC
Enforcement of Fire Codes: State Fire Marshal, Department of Public Safety, Division of Fire Protection, 1010 Commonwealth Avenue, Boston, MA 02215; 617–566–4500
Life Safety Code: 1976 NFPA 101 referenced in building code for certain institutional uses
Enforcement of Life Safety Codes: none

Michigan

Building Code: 1987 NBC
Enforcement of Building Codes: local or Chief Inspector, Building Division Bureau of Construction Codes, Department of Labor, State Secondary Complex, 7150 Harris Drive, PO Box 30015, Lansing, MI 48909; 517–322–1075
Fire Prevention Code: none statewide, promulgated locally
Enforcement of Fire Codes: Captain, Department of State Police, Fire Marshall Division, 7150 Harris Drive, Lansing, MI 48913; 517–322–1924
Life Safety Code: none statewide, promulgated locally
Enforcement of Life Safety Codes: same as fire codes

Minnesota

Building Code: 1985 UBC
Enforcement of Building Codes: certified local building officials or Director, Division of Building Codes & Standards, 408 Metro Square Building, St. Paul, MN 55101; 612–296–4627
Fire Prevention Code: 1982 UFC
Enforcement of Fire Codes: State Fire Marshal, Market House, 289 E. 5th Street, St. Paul, MN 55101; 612–296–7643
Life Safety Code: 1982 UFC
Enforcement of Life Safety Codes: same as fire codes

Mississippi

Building Code: none
Enforcement of Building Codes: none
Fire Prevention Code: none
Enforcement of Fire Codes: none
Life Safety Code: none
Enforcement of Life Safety Codes: none

Missouri

Building Code: none statewide, promulgated locally
Enforcement of Building Codes: local
Fire Prevention Code: none statewide, promulgated locally
Enforcement of Fire Codes: local
Life Safety Code: none statewide, promulgated locally
Enforcement of Life Safety Codes: local

Montana

Building Code: 1988 UBC
Enforcement of Building Codes: local or Bureau Chief, Building Codes Bureau,
    Department of Commerce, Capital Station, Helena, MT 59601; 406–444–3933
Fire Prevention Code: 1988 UFC
Enforcement of Fire Codes: local or State Fire Marshal, State Fire Marshal Bureau,
    1409 Helena Avenue, Helena, MT 59601; 406–444–2050
Life Safety Code: none
Enforcement of Life Safety Codes: none

Nebraska

Building Code: none statewide, promulgated locally
Enforcement of Building Codes: local
Fire Prevention Code: none
Enforcement of Fire Codes: none
Life Safety Code: 1985 NFPA 101
Enforcement of Life Safety Codes: State Fire Marshal, PO Box 94677, Lincoln, NE
    68509; 402–471–2027

Nevada

Building Code: none statewide, promulgated locally
Enforcement of Building Codes: local

Fire Prevention Code: none statewide, promulgated locally
Enforcement of Fire Codes: local
Life Safety Code: none statewide, promulgated locally
Enforcement of Life Safety Codes: local

New Hampshire

Building Code: 1981 NBC for new public buildings only
Enforcement of Building Codes: local
Fire Prevention Code: 1981 NFC for new public buildings only, 1981 NFPA standards
Enforcement of Fire Codes: State Fire Marshal, Department of Safety, James H. Hayes Building, Concord, NH 03305; 603–271–3294
Life Safety Code: 1981 NFPA 101
Enforcement of Life Safety Codes: same as fire codes

New Jersey

Building Code: 1987 NBC with 1988 supplement
Enforcement of Building Codes: Director, Division of Housing and Community Development, 101 S Broad Street, Trenton, NJ 08625; 609–292–7899
Fire Prevention Code: 1985 NJ Uniform Fire Code (NFC)
Enforcement of Fire Codes: same as building codes
Life Safety Code: 1985 NJ Uniform Fire Code (NFC)
Enforcement of Life Safety Codes: same as building codes
  Note: The New Jersey building code is divided into subcodes. Within the building code is the fire protection subcode, which applies to new construction. The NJ Uniform Fire Code, based on BOCA amendments, covers fire and life safety for existing buildings.

New Mexico

Building Code: 1988 UBC
Enforcement of Building Codes: Bureau Chief, General Construction Bureau, Bataan Memorial Building, Santa Fe, NM 87503; 505–827–6261
Fire Prevention Code: UBC by reference
Enforcement of Fire Codes: State Fire Marshal, PERA Building, PO Drawer 1269, Santa Fe, NM 87504, 505–827–4500 x29
Life Safety Code: same as fire prevention code
Enforcement of Life Safety Codes: same as fire codes

New York

Building Code: 1988 Uniform Fire Prevention & Building Code
Enforcement of Building Codes: local or Director, Codes Division, Department of State, 162 Washington Avenue, Albany, NY 12231; 518–474–4073

Fire Prevention Code: same as building code
Enforcement of Fire Codes: same as building code
Life Safety Code: same as building code
Enforcement of Life Safety Codes: same as building codes

North Carolina

Building Code: 1978 General Construction Code (Volume 1)
Enforcement of Building Codes: local or Deputy Commissioner, Department of Insur-
   ance, 410 N. Boylan Avenue, Raleigh, NC 27603; 919–733–3901
Fire Prevention Code: none statewide, promulgated locally
Enforcement of Fire Codes: local
Life Safety Code: none statewide, promulgated locally
Enforcement of Life Safety Codes: local
   Note: Pertaining to the Life Safety Code, the state uses NFPA 101 as reference
      standard only. NFPA 101 conflicts with the state building code, therefore it
      cannot be adopted. If a locality adopts NFPA 101, the state building code is
      final authority.

North Dakota

Building Code: 1985 UBC
Enforcement of Building Codes: local
Fire Prevention Code: 1985 UFC and 1985 UBC
Enforcement of Fire Codes: local
Life Safety Code: reference 1985 NFPA 101
Enforcement of Life Safety Codes: local

Ohio

Building Code: 1987 NBC
Enforcement of Building Codes: local or Chief, Division of Factory Building Inspec-
   tion, 2323 W. 5th Avenue, PO Box 825, Columbus, OH 43216; 614–644–2622
Fire Prevention Code: 1984 NPC
Enforcement of Fire Codes: local or State Fire Marshal, 8895 E. Main Street, Rey-
   noldsburg, OH 43068; 614–864–5510
Life Safety Code: none
Enforcement of Life Safety Codes: none
   Note: Pertaining to fire prevention, local building departments must be certified by
      the state before they can approve plans and make inspections. Pertaining to
      life safety, some local jurisdictions have adopted NFPA 101.

Oklahoma

Building Code: none statewide, promulgated locally
Enforcement of Building Codes: local

Fire Prevention Code: none statewide, promulgated locally
Enforcement of Fire Codes: local
Life Safety Code: none statewide, promulgated locally
Enforcement of Life Safety Codes: local
   Note: Pertaining to fire prevention and life safety, the fire marshal's office does plan reviews and inspections. It applies 1988 NFPA 101 and 1987 BOCA National Building Code.

Oregon

Building Code: 1988 UBC
Enforcement of Building Codes: local *or* Administrator, Building Codes Agency, 1535 Edgewater NW, Salem, OR 97310; 503–378–3176
Fire Prevention Code: 1985 UFC
Enforcement of Fire Codes: local, this has been updated to 1988 UFC effective January 1990
Life Safety Code: none, except NFPA 101 applies to hospitals and nursing homes only
Enforcement of Life Safety Codes: none

Pennsylvania

Building Code: none statewide, promulgated locally
Enforcement of Building Codes: local
Fire Prevention Code: 1986 Fire and Panic Regulations; Philadelphia, Pittsburgh, and Scranton are exempt
Enforcement of Fire Codes: Director, Bureau of Occupational and Industrial Safety, Department of Labor and Industry, Room 1514, Labor and Industry Building, Harrisburg, PA 17120; 717–787–3806
Life Safety Code: same as fire prevention code
Enforcement of Life Safety Codes: same as fire codes

Rhode Island

Building Code: 1987 NBC
Enforcement of Building Codes: local *or* Commissioner, State Building Commission, 610 Mount Pleasant Avenue, Building 2, Providence, RI 02908; 401–277–3033
Fire Prevention Code: 1988 Rhode Island Fire Safety Code
Enforcement of Fire Codes: local *or* State Fire Marshal, Division of Fire Safety, 1270 Mineral Spring Avenue, Providence, RI 02904; 401–277–2335
Life Safety Code: contact the state fire marshal for specific applications, various editions may apply
Enforcement of Life Safety Codes: see life safety code

South Carolina

Building Code: none, however 1988 SBC with state amendments are mandatory for state buildings, voluntary for locals. Locals may amend codes, with state approval, for physical or climatical reasons.
Enforcement of Building Codes: none
Fire Prevention Code: none, however 1988 Standard fire prevention codes with 1987 state amendments are mandatory for state buildings, voluntary for locals.
Enforcement of Fire Codes: none
Life Safety Code: none
Enforcement of Life Safety Codes: none

South Dakota

Building Code: none
Enforcement of Building Codes: none
Fire Prevention Code: none statewide, promulgated locally
Enforcement of Fire Codes: local
Life Safety Code: 1985 NFPA 101 and 1985 UFC
Enforcement of Life Safety Codes: local or Fire Marshal, 118 W. Capitol Pierre, SD 57501; 605–773–3562

Tennessee

Building Code: 1988 SBC
Enforcement of Building Codes: local, however local jurisdictions may obtain exemptions from the state code; approximately twenty larger cities are exempt.
Fire Prevention Code: 1988 NFC volumes 1–12
Enforcement of Fire Codes: local
Life Safety Code: 1988 NFPA 101
Enforcement of Life Safety Codes: local
  Note: Educational and state-owned buildings should contact the Director, Fire Safety Section, 1808 W. End Building, Nashville, TN 37219; 615–741–7162

Texas

Building Code: none state wide, promulgated locally
Enforcement of Building Codes: local, however for industrialized buildings 1988 UBC and 1988 SBC are mandatory
Fire Prevention Code: various NFPA standards
Enforcement of Fire Codes: State Fire Marshal, State Board of Insurance, 1110 San Jacinto, Austin, TX 78701; 512–322–3550
Life Safety Code: 1985 NFPA 101

Enforcement of Life Safety Codes: Associate Commissioner, Department of Health, 1100 W. 49th Street, Austin, TX 78756; 512–458–7111

Note: Pertaining to fire prevention, the Board of Insurance regulates installers of sprinklers, fire extinguishers, and fire alarms. Pertaining to life safety in health care facilities, contact the administration at 512–458–7531. Pertaining to life safety in long-term care facilities, contact the administration at 512–458–1346.

Utah

Building Code: 1988 UBC
Enforcement of Building Codes: local, however the Uniform Standards Commission determines if proposed local amendments will be adopted or rejected and if the amendment will be adopted statewide or by local jurisdiction
Fire Prevention Code: none statewide, promulgated locally
Enforcement of Fire Codes: local
Life Safety Code: none statewide, promulgated locally
Enforcement of Life Safety Codes: local

Note: Pertaining to fire prevention, the 1988 UFC applies to state-owned buildings, day care centers, hospitals, jails, and educational buildings. Any questions, contact the State Fire Marshal, 4501 S. 2700 West, Salt Lake City, UT 84119; 801–965–4353. Pertaining to life safety, the 1988 NFPA chapters 10–15 apply to state-owned buildings, day care centers, hospitals, jails, and educational buildings. The state's written code applies to residential care facilities. Contact the State Fire Marshal, 4501 S. 2700 West, Salt Lake City, UT 84119; 801–965–4353.

Vermont

Building Code: 1981 NBC
Enforcement of Building Codes: Director, Fire Prevention Bureau, Department of Labor and Industry, 120 State Street, Montpelier, VT 05602; 802–828–2106
Fire Prevention Code: 1981 NFC
Enforcement of Fire Codes: same as building code
Life Safety Code: none
Enforcement of Life Safety Codes: none

Note: The building code is also known as the 1983 Vermont Fire Prevention and Building Code. Pertaining to life safety, NFPA 101, chapters 12, 13, 14, and 15, with state amendments, are used for medicaid and correctional facilities. Only medicaid and correctional facilities have mandatory life safety codes that are enforced by the director of fire prevention. Address: Department of Labor and Industry, 120 State St. Montpelier, VT 05608; 802–828–2106.

Virginia

Building Code: 1987 NBC
Enforcement of Building Codes: local or Associate Director, Code Enforcement Office, Division of Building Regulatory Services, 205 N. 4th Street, Richmond, VA 23219; 804–786–5041
Fire Prevention Code: 1987 NFC
Enforcement of Fire Codes: local or Chief Fire Marshal, Office of State Fire Marshal, 205 N. 4th Street, Richmond, VA 23219; 804–786–4751
Life Safety Code: none, however, 1981 NFPA 101 is enforced for medicaid and medicare facilities
Enforcement of Life Safety Codes: none

Washington

Building Code: 1988 UBC
Enforcement of Building Codes: local, however locals may not amend for 1 to 4-unit residential buildings
Fire Prevention Code: 1988 UFC
Enforcement of Fire Codes: local
Life Safety Code: see building and fire codes
Enforcement of Life Safety Codes: see building and fire codes

West Virginia

Building Code: 1987 NBC
Enforcement of Building Codes: local or State Fire Marshal, State Capitol, 2100 Washington Street E, Charleston, WV 25305; 304–348–2191
Fire Prevention Code: 1983 NFC
Enforcement of Fire Codes: same as building codes
Life Safety Code: 1981 NFPA 101
Enforcement of Life Safety Codes: same as building codes

Note: Pertaining to fire prevention, they anticipate adoption of 1987 standards, including NFPA 101.

Wisconsin

Building Code: 1986 State Building Code
Enforcement of Building Codes: local or Administrator, Division of Safety and Buildings, Department of Labor, Industry and Human Relations, PO Box 7969, Madison, WI 53707; 608–266–1816
Fire Prevention Code: 1989 State Fire Prevention Code
Enforcement of Fire Codes: local
Life Safety Code: none
Enforcement of Life Safety Codes: none

Wyoming

Building Code: 1988 UBC with 1989 supplement
Enforcement of Building Codes: State Fire Marshal, Department of Fire Prevention, Herschler Building, Cheyenne, WY 82002; 307–777–7288
Fire Prevention Code: 1988 UFC with 1989 supplement
Enforcement of Fire Codes: same as building codes
Life Safety Code: none
Enforcement of Life Safety Codes: none

District of Columbia

Building Code: 1984 NBC with 1985 supplement
Enforcement of Building Codes: Administrator, Building and Land Use, Regulation Administration, 614 H Street NW, Room 312, Washington, DC 20001; 202–727–7340
Fire Prevention Code: 1984 NEC with 1985 supplement
Enforcement of Fire Codes: Fire Marshal, DC Fire Prevention Division, 613 G Street NW, Room 810, Washington, DC 20001; 202–745–2345
Life Safety Code: none
Enforcement of Life Safety Codes: none

Puerto Rico

Building Code: 1968 Building Code
Enforcement of Building Codes: Administrator, Regulations and Permits Administration, Minillas Government Center, Santurce, PR 00904; 809–721–8282
Fire Prevention Code: 1968 Building Code
Enforcement of Fire Codes: Fire Chief, Fire Service of Puerto Rico, Box 13325, Santurce, PR 00908; 809–723–5776
Life Safety Code: 1968 Building Code
Enforcement of Life Safety Codes: same as building codes

Virgin Islands

Building Code: 1972 VI Building Code; legislation is pending to adopt the UBC
Enforcement of Building Codes: Commissioner, Department of Planning and Natural Resources, Nisky Center Suite 231, St. Thomas, VI 00801; 809–774–3320
Fire Prevention Code: 1985 NFC and 1985 NFPA standards
Enforcement of Fire Codes: District Fire Marshal, 129 and 133 Chinnery Building, Subbase, St. Thomas, VI 00802; 809–774–7610
Life Safety Code: 1985 NFPA 101
Enforcement of Life Safety Codes: same as fire codes

SUMMARY

The proper use and enforcement of codes depends solely on the state or local municipality. Although codes conflict, they primarily try to ensure a safer facility. As the FSD, you should realize that these codes are not to punish you or your facility, but are to ensure that corrective measures are taken so that your facility complies with the codes.

> A common element of the worst life loss fires of recent years has been the involvement of properties that are on the fringes of code enforcement. The 1965 Searcy, AR, missile silo fire and the 1977 Columbia, TN, county jail fire both involved government properties, which typically are outside the jurisdiction of local fire departments. The 1977 Beverly Hills Supper Club fire, the 1963 Golden Age Nursing Home fire, and the 1972 silver mine fire in Idaho all occurred in small communities where resources for fire code enforcement typically are very limited.[29]

The NFPA building and life safety codes are written and enforced so that facilities are safer for individuals who work, live, or visit that building.

"Although all of these laws, ordinances and codes are necessary for the helpful development of the community, their effectiveness will be greatly diminished unless an adequate, well-qualified administration is provided. The administrative staff should be carefully selected and well organized to serve the public effectively and efficiently."[30]

No matter how codes are interpreted or enforced it is still the responsibility of competent municipalities to select qualified candidates to oversee the enforcement of these regulations. The FSD should remember that errors of interpretation and enforcement are made on all levels. If you feel that an error has been made, contact the agency in writing and seek a clarification of that interpretation. Never assume that any agency official is correct. If a summons or a violation is issued to you, accept it, but advise the official that you are accepting it under protest. If the summons requires your signature, sign it, but also note that you are signing it under protest.

Codes and their enforcement will continue to become more complex in the 1990s. But, remember, they ensure the continued protection of life and property.

7

Architecture, Security, and Fire Prevention

When builders decide to erect or rebuild buildings they hire a competent architect to design the structure. The builder has the architect submit the proposed drawings to the building department for approval. The building department reviews the architect's drawings to see if they are in compliance with the various codes. Once the building department approves the drawings, construction may begin. Simple enough? Sounds easy, however there is more to this scenario, as we will see throughout this chapter.

Today's architect/engineers (A/Es) consider a multitude of ideas before designing a building. The location, type of construction, and operation require architects to incorporate fire safety and security into their overall design. Therefore, aside from designing a building that is visually pleasing, the architect must also incorporate fire safety and security features into the facility.

When the architect begins a project, she is the quarterback of that job. (See Figure 7–1) She is responsible for complying with all building codes and regulations. Mechanical, electrical, plumbing, and elevator engineers report to the architect. Her signature on the drawings indicates that the building's design is in compliance with building codes and regulations. The architect will also consult with the building and fire departments. Why, then, in most instances, isn't a security/fire safety professional's advice and input sought?

The security/fire safety professional can give insight into the alarm and detection systems required by the new facility. The security/fire safety professional can aide the architect in the areas of egresses, alarm systems, lighting, fire safety, fire prevention, sprinkler systems, and new products on the market that would enhance in the building design.

According to William Collins from the AIA, "with regard to fire safety, an architect would first decide on the construction classification of the building based upon its occupancy, size and location (in or outside of the fire limits) when designing a new building. During the design phase of the building, many items will be considered with respect to fire safety. They would be:

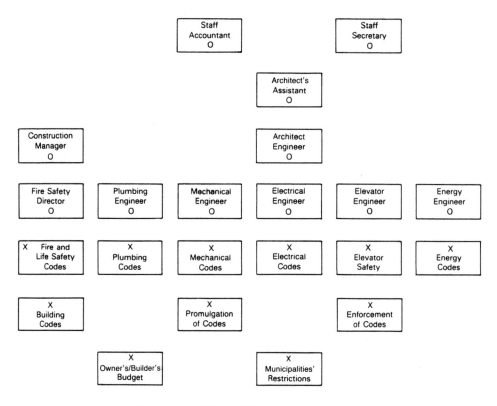

Figure 7–1 The architect as the quarterback.

1. means of egress from all points of the building
2. fire separations between tenants
3. fire areas within the building
4. fire rating of materials used
5. sprinkler systems
6. emergency lighting
7. location of stair towers

It would be crucial at the design stage to incorporate all of these items into an architect's plan."[31]

Collins goes on to add that architects emphasize building safety as mandated by the applicable codes. As for the design of a building versus the cost of the building, be aware that many architects may become overzealous and will design a building that exceeds the owner's budget.

Many security and fire safety professionals with no background in architecture or design concepts have heard the term *fire wall* at some stage of their learning process. Many security and fire safety professionals ask what determines when to construct a 2-hour or 4-hour fire wall in a facility.

The answer is a simple one. The occupancy or use, size, and construction classification will dictate the minimum resistance requirements of a building. (See Table 7–1; extracted from *The New York State Fire Prevention and Building Code*.)

Nowadays, due to liability, architects are generally concerned with fire safety as well as implementing innovative ideas. Changes in the various codes have addressed changes in building designs. Atriums, for example, a major design element in many buildings, now require *smoke evacuation systems*. The smoke evacuation system requires ventilation and removal of smoke, gases, and heat from a building. Ventilation of building spaces accomplishes the following functions:

1. Protection of life by removing or diverting toxic gases and gases from locations where building occupants must find temporary refuge.
2. Improvement of the environment in the vicinity of the fire by removal of smoke and heat. This enables fire fighters to advance close to the fire to extinguish it with a minimum of time, water, and damage.
3. Control of the spread or direction of fire by setting up air currents that cause the fire to move in a desired direction. In this way occupants or valuable property can be more readily protected.
4. Provision of a release for unburned, combustible gases before they acquire a flammable mixture, thus avoiding a backdraft or smoke explosion[32]

Another innovative trend in architecture and fire safety design is architects using *fire resistant glass*. The glass is clear and aesthetically pleasing in design and, in the event of a fire, will not give off toxic gases. Many of today's structures are using fire resistant glass. The security/fire safety director should be aware of this product and know that it is acceptable to use in the design of a building.

WORKING WITH THE ARCHITECT/ENGINEER

Any experienced security or fire safety director wants the architect to design functional, simple safety and security features into a building. However, not all architects seek the advice of the security or fire safety director. In all fairness to architects, the position of security or fire safety director has most often not even been filled or even budgeted for when the original designs are submitted. However, if the security or fire safety director's input is sought prior to the design phase, his primary objectives should be to eliminate criticality and vulnerability.

Criticality are highly sensitive policies, procedures, and systems in which the threat of open attack or damage to corporate operations is considerable. *Vulnerability* is weak or inadequate policies, procedures, and sys-

Table 7–1a Minimum fire-resistance requirements of structural elements
(By types of construction: fire-resistance rating in hours)

Structural Element	Type 1 (Fire-resistive)	Type 2 (Non-combustible)		Type 3 (Heavy timber)	Type 4 (Ordinary)		Type 5 (Wood frame)	
		2a	2b		4a	4b	5a	5b
Exterior:								
Bearing walls	3	2	nc	2	2	2	¾	c
Nonbearing walls[2]	2	2	nc	2	2	2	¾	c
Panel and curtain walls	¾	¾	nc					
Party walls[3]	2	2	2	3	2	2	2	2
Interior:								
Fire walls[4]		2						
Bearing walls or partitions	3	3	2	2	3	2	2	2
Partitions enclosing stairways, hoistways, shafts, other vertical openings and corridors	2	2	nc	2	¾	c	¾	c
Construction separating tenant spaces	1	1	¾	¾	¾	¾	¾	¾

Construction Classification[1]

Columns, beams, girders and trusses (other than roof trusses); supporting more than 1 floor or 1 floor and a roof supporting 1 floor or a	3	2	nc	3/4	3/4	c	3/4	c
roof	2	3/4	nc	3/4	3/4	c	3/4	c
Floor construction, including beams	2	1	nc	3/4	3/4	c	3/4	c
Roof construction, including purlins, beams and roof trusses	1	3/4	nc	3/4	3/4	c	3/4	c

Extracted from: The New York State Fire Prevention & Building Code Book.

[1] For classification of building by type of construction, see Part 704.

[2] For exceptions, see 738.3(a)(2) and 738.3(b)(2).

[3] Party walls shall comply with section 738.8.

[4] Fire walls shall comply with section 739.2.

[5] In buildings not more than three stories in height, and with not more than eight dwelling units within a fire area, one hour in type 1 construction; 3/4 hour in type 2, 3 and 4 construction. See 739.4(d)(7).

[6] If every part of noncombustible roof truss is more than 15 feet above floor next below, protection of the roof truss is not required. Roof construction shall be of noncombustible material, but is not required to have any rating.

[7] In buildings of types 2, 3 and 4 construction, more than three stories in height, the floor of the lowest story and all construction below shall be type 1.

[8] 3/4 hour when separating tenant spaces, and for floor construction over a cellar or basement.

[9] For atrium exceptions, see Part 743.

Table 7–1b Height and fire area for buildings of groups C1, C2, C3 and C4 occupancy
See sections 706.4(e) and 705.4(f) for increased height or fire area; see section 774.4 for sprinkler requirements.

| In Stories | In Feet | Type 1 (Fire-resistive) | | Type 2 (Non-combustible) | | Type 3 (Heavy timber) | Type 4 (Ordinary) | | Type 5 (Wood frame)[1] | |
		1a	1b	2a	2b	3	4a	4b	5a	5b
				LOW HAZARD—C1, C3.1, C4.1						
1	un	un	un	un	18,0000	21,000	18,000	12,000	9,000	6,000
2	40	un	un	21,000	15,000	18,000	16,000	9,000	6,000	3,000
3	55	un	un	16,000	np	15,000	12,000	6,000	np	np
4	70	un	un	15,000	np	12,000	9,000	np	np	np
5	85	un	un	12,000	np	np	np	np	np	np
6	100	un	un	np	np	np	np	np	np	np
More than 6	More than 100	un	un	np	np	np	np	np	np	np
				MODERATE HAZARD—C2, C3.2, C4.2						
1	un	un	un	18,000	15,000	15,000	15,000	8,000	8,000	4,000
2	40	un	30,000	15,000	12,000	12,000	12,000	6,000	4,000	2,000

Basic Fire Area by Construction Classification in Square Feet

Stories	Height									
3	55	un	28,000	12.000	np	9,000	9,000	4,000	3,000	2,000
4	70	un	26,000	10,000	np	np	np	np	np	np
5	85	un	24,000	np	np	np	np	np	np	np
6	100	un	22,000	np	np	np	np	np	np	np
More than 8	More than 100	un	np	np	np	np	np	np	np	np
HIGH HAZARD—C3.3, C4.3										
1	un	24,000	15,000	8,000	6,000	6,000	6,000	4,000	3,000	2,000
2	40	23,000	14,000	7,000	5,000	5,000	5,000	np	np	np
3	55	22,000	13,000	8,000	np	np	np	np	np	np
4	70	21,000	12,000	np	np	np	np	np	np	np
5	85	20,000	np	np	np	np	np	np	np	np
More than 5	More than 85	np	np	np	np	np	np	np	np	np

[1] Not permitted within fire limits.

[2] For aircraft hangers, basic fire areas may be increased 26 percent.

[3] Fire area of a one-story building may be unlimited provided that the building is located outside the fire limits, has open unobstructed space on all sides accessible for firefighting, as set forth in 706.4(e)(1) and such space shall be at least 50 feet wide. Heat banking areas and an automatic sprinkler system as set forth in sections 771.4(e) and 1060.4 respectively, shall be provided except that in a building group C3.1 occupancy of type 2b construction used for the manufacture or processing of noncombustible products, where the distance from the floor to the lowest point of the roof structure is 20 feet or more, heat banking areas and an automatic sprinkler system are not required.

[4] See Part 774 for fire protection equipment requirements.

tems that seriously affect the safety and security of personnel, assets, and profitability.

Too often, the FSD is not made aware of a problem in the architect's design until after the fact. However, once the FSD is made aware of a potential problem, she must analyze the criticality of the problem and determine how to remedy it.

The security director who has conducted numerous building surveys will look for vulnerable areas in the building design from a security standpoint. However, how vulnerable is the property from a fire safety standpoint? Remember, you are responsible for ensuring fire safety as well as security in the building design.

Communicating Resolutions

Most organizations complain how vulnerable their organization is to crime and how billions of dollars are lost daily throughout the United States due to employee theft. However, organizations never consider how much that can be lost through an employee's carelessness regarding fire safety. Organizations are probably more vulnerable to their company closing due to a fire than through merchandise thefts.

The fine art of diplomatically communicating your ideas to A/Es will help if they are seeking input. If A/Es are not seeking input, your ideas and suggestions will fall on deaf ears. Most security/fire safety directors' primary purpose is to protect their organizations' assets. If the organization continually seeks your advice, why would an A/E not seek your expertise regarding security and fire safety decisions? Many times, the FSD refuses to listen to what the A/E has to say. It is crucial that you remember that the A/E is responsible for all phases of building design and construction. Security system design and construction has much in common with other types of design and construction. Therefore, the project management techniques and the design documentation generated by professional fields such as civil, mechanical, or electrical engineering can offer guidelines for the development of security systems.[33]

Thomas J. Whittle, PE, a member of the Architect-Engineer Subcommittee of the American Society for Industrial Security's Standing Committee on Physical Security outlined a six-phase sequence that security directors should follow when becoming involved with design and construction of a project. (See Table 7–2)

The security/FSD must realize how she fits within the A/E's overall hierarchy. Most A/Es develop a project management plan at the inception of the building project. Information pertaining to that project, including security and fire safety, are referenced in the plan. The project management plan establishes the groundwork for conducting business throughout the project.

Table 7-2 Design documentation format

Design Stage	Design Team Activities
Study and report phase	Develop functional requirements Perform an economic analysis Create schematic layouts Develop conceptual design with alternatives Validate primary design concept Present concepts to client
Preliminary design phase	Create total project scoping Develop design documentation —create preliminary drawings —outline specifications —create design analysis —create cost estimate Present design to client
Final design phase	Complete design documentation Develop supplemental data —bid forms —invitations for bids —instructions to bidders Present final design to client
Bidding and negotiation phase	Attend prebid conference Provide administrative and technical support to contract personnel
Construction phase	Evaluate construction security Assist with contract administration Perform construction inspection Perform system test and acceptance routines Evaluate equipment substitutions Create systems manuals
Operational phase	Provide administrative and technical support to the user Assist in training Develop plans and procedures[32]

Source: Thomas J. Whittle, "The Phases of Partnership." *Security Management,* 34 (April 1990):39.

Security and fire safety should be an important phase of the project management plan. However, the A/E has the overall authority to incorporate security and fire safety when he believes it's necessary. The security/FSD must remind the A/E of the necessity of specific security and fire safety strategies, equipment, and systems. If new systems are to be incorporated in the facility, the A/E should be advised of this and provided with a detailed outline of the system.

Study and Report Phase

The study and report phase is when the foundation of the project is developed and design, costs, and concepts are created. The building site will be surveyed by the A/E and surveys will be undertaken to define the builder's criticality and vulnerability.

Schematic layouts and conceptual designs are created during this phase. These are then issued to the owner to see if they meet her requirements. The cost is also projected during this phase. The projected costs serve as an estimate of the total cost of the project and allows the A/E to incorporate other expenditures that the owner may have overlooked or underestimated in his original budget.

As the security/FSD, you should be involved during the study and report phase. Security and fire safety systems and programs should be discussed and incorporated into the design and budget. Provide the A/E and owner with professional input and recommendations regarding security and fire safety concepts. Your predesigned fire operation and fire safety plans should be discussed at this point. Your concerns regarding security, fire, and communications systems should be brought to the attention of the A/E. You are responsible for making the A/E aware of potential problems he may face in design concepts regarding these areas.

It is important for your recommendations to be written in the simplest terms. For example, many years ago, a colleague of mine who is a director of fire safety and security wrote a memo to his staff regarding excessive lights being left on in a high-rise office building. He wrote, "due to the severe energy crisis we Americans are facing, it is important for each of us to uphold our American civic duty. Between the hours of dusk and dawn, the illumination in our office building remains on. The wattage we are wasting is unnecessary. Therefore, I duly propose, effective as of tomorrow, between dusk and dawn, when offices are unoccupied, the illumination of that office be extinguished." It would have been easier to simply say, "please shut off all interior office lights when you leave for the day." Do not insult the intelligence of the A/E or owner by saying something just to be heard. Make sure your remarks are important and are justified.

Preliminary Design Phase

Beginning at the preliminary design phase, the A/E creates precursory drawings and specifications for the project in outline form. The design documentation created by the A/E is usually extensive and, at this stage, is incomplete. The A/E will go back to his drawing board many times prior to the final drawing being accepted and approved.

The expertise of the A/E in the area of finance must be applied at this stage. Cost analysis is usually more accurate at this stage when quotes on materials, hardware, and labor are fixed and the budget is structured. The owner will communicate daily with the A/E to discuss designs and plans.

The security/FSD should be involved throughout this stage as well. When security, alarm, communication, and fire systems are discussed, the information requested from you by the A/E is greatly needed. Designs and budgets cannot be completed until the A/E receives a determination from a competent security/fire safety professional as to the systems that should be installed to protect the facility from crime and fire.

All projects are incorporated through long-range planning. The A/E will be on her schedule and the owner will watch this schedule carefully as business executives do not like to overspend. If schedules are extended, it usually costs money. You can save the owner money by meeting with the A/E and discussing the role that security and fire safety will play in the A/E's overall plan.

The fire operation plan, fire safety plan, security procedures, and loss prevention programs must be described to the A/E. Provide the A/E with a list of goals and how you hope to achieve them. Research security and fire safety systems and discuss these systems with the A/E. Due to the many sophisticated security and fire safety systems on the market today, the A/E is probably unaware of how much should be allotted in the budget for security and fire safety. If the recommendation is presented by the security/FSD in a professional manner, he will be one step closer to achieving his overall purpose—the protection of life and property.

Final Design Phase

Frustrating, infuriating, and complicated are just a few of the words that describe the final design phase. The A/E, her support staff, the owner, and the security/FSD are either in complete agreement, partial agreement, or complete disagreement regarding the ideas, trends, concepts, and costs discussed during the first two design phases.

In the preliminary design phase, the owner has several meetings with the A/E and the support staff. If communication was a problem in the preliminary design phase, by the final design phase you will have a serious problem with which to deal. As the security/FSD director, your job is done at this stage of design. Patience is often the best asset to have during this phase. During the final design phase, the A/E may need to rethink her game plan due to improper or inadequate research in nonsecurity fields. At this point, it is best to sit back and let the A/E handle the situation.

In the final design phase, the A/E will have completed the design diagrams and finalized the projected cost estimate. The security/FSD should now begin asking questions regarding bid forms, potential vendors, and requests for proposals (RFPs). If the A/E has decided to incorporate a security and/or fire system, this is the time to begin selecting vendors.

Major changes in the A/E's plan is uncommon in this phase, unless, of course, the owner requests that specific changes be made. Another reason changes may be made are due to the security/FSD, or other support staff members miscalculating estimated costs or failing to effectively research specified areas of the project.

Bidding and Negotiation Phase

The A/E will seek the input of his support staff regarding appropriate vendors and how they should be selected. As the security/FSD, the preferred method of making this selection is to be present when the bidding and negotiation phase of the project begins to ensure that the bidding process is correctly executed. The FSD may also be able to begin financial background checks on proposed vendors. Background checks will ensure that no collusion has taken place between the A/E, the support staff, and proposed vendors.The A/E and other support staff may regard these checks as an intrusion. However, if the checks are professionally executed, no one should feel insulted. It is vital to remember that the A/E may be unaware of any collusion between the support staff and the vendors.

Construction Phase

Most construction companies think that the proprietary and contract guard agencies are insurance companies as well. More materials disappear at the end of a construction job and the foreman eventually comes to the security/FSD to demand payment for missing materials. This is not how security and fire safety professionals operate. If faced with this situation, my reply to them is to advise the foreman to notify the local police and the insurance company of the thefts.

When a facility is under construction, the responsibilities of the security/fire safety department increase. The security/FSD should advise the A/E of the areas of security and fire safety that will be affected during the construction phase. *Target hardening*, which is associated with surveying a physical property and strengthening security practices, should be enacted. (See Figure 7–2) *Areas to strengthen* during target hardening include

access control	increased fire tours
egress control	lighting control
package control	communication procedures
identification procedures	manpower

The A/E or the owner may not be aware of the activities needed to effectively secure and safeguard the property during the construction phase. If designs or plans are changed during the construction phase, the security/FSD should be advised. Notification will assist him with his long-range plans.

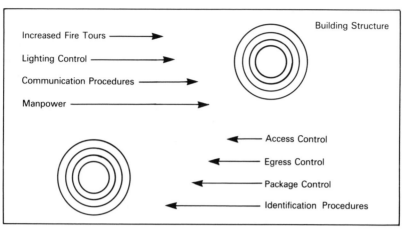

Increased Fire Tours

Lighting Control

Communication Procedures

Manpower

Building Structure

Access Control

Egress Control

Package Control

Identification Procedures

Figure 7–2 Target hardening. By strengthening these areas during the construction phase, your chances of creating and establishing tighter security and fire safety procedures increase. Target these areas as a starting point. Inform the A/E that these areas must be incorporated into the planning stage.

Operational Phase

The ultimate goal of the A/E and her support staff is to construct the facility on time, under budget, and operating effectively. Unfortunately, problems develop, costs increase, and systems fail. To compensate for these unforeseen problems, the A/E must assist in completing all financial reports and participate in the training and development of all operating systems.

From your perspective, the operational phase is crucial to your department. The A/E and the support staff's primary concern is the overall operation of the facility and your department is a small segment of the finished product. If your department becomes one less area of concern for the A/E and the owner, you have executed your job correctly and professionally.

Thomas J. Whittle, PE, author of *The Phases of Partnership*, emphasizes the importance of A/E and the security/FSD working in conjunction with one another. Whittle states, "too often, poorly prepared security directors have allowed ill-conceived design concepts to be approved . . ."[34] The reality of this statement is that most security directors are not fire safety directors. They may be astute in law enforcement and security concepts, but they do not have a single notion of how to design or formulate a fire safety plan in conjunction with A/Es.

The security/FSD is not expected to be able to design or read blueprints. However, he should be able to speak in terms of security and fire safety concepts. If asked his opinion regarding security or fire safety systems, the FSD should be able to discuss the systems in terms of operational capabilities. If the A/E or the support staff ask your opinion regarding these systems and then show you blueprints, ask questions. Do not pretend you know how to read them.

FIRE-SAFE BUILDINGS/TESTING METHODS

Every architect wants to design a spectacular building that is also a fire-safe structure. Nowadays, design professionals concentrate more and more on fire protection engineering. The skills and techniques incorporated in designing a building take time, planning, and money.

A/Es who use the services of fire protection engineers want to create and incorporate sound fire protection and fire prevention systems into the structure's design. As the security/FSD, you need to ask about four specific areas:

1. Hazards to life in structures
2. Water and water equipment for fire protection
3. Water-based extinguishing systems
4. Fire suppression agents and systems

These four areas are a good starting point for the security/FSD to begin asking questions. Remember, your primary concern is the protection of life and property. In regard to hazards to life in structures, you may want to ask, Have we complied with the various codes to make this facility as safe as possible? Are there any hazards we may have overlooked in this facility? Have we designed an effective fire detection and suppression system suitable for this facility?

The next two areas deal with water systems to fight a fire and/or prevent it from spreading. Many times, a fire will spread even though you have taken every conceivable precaution. Therefore, you may want to ask, Is there a sufficient number of sprinklers installed throughout the facility? If we do not have a sprinkler system, what is our main system to combat a fire? Does the facility have adequate water pressure? Aside from water-extinguishing systems, what other suppression systems are we using? How effective are these systems compared to a water-based extinguishing system?

The fourth area of fire suppression and containment is vital. After all previous questions have been discussed and answered, all parties involved in building the facility should be aware of and understand other fire suppression systems and combating agents. By reading and networking, you should be able to answer a multitude of questions regarding fire systems. The NFPA is an excellent starting point to indicate alternative fire systems. Any questions that may arise pertaining to fire systems should be checked with the NFPA. In some instances, when answers cannot be obtained immediately, it is beneficial to write to the NFPA and present your question, and ask them for a written answer to the issue.

THE X-FACTOR

In Chapter 1, I spoke of the X-factor regarding human behavior and fire. Remember, no two fires are alike and no two persons react the same during a fire. Interesting enough, the X-factor plays an integral role with architects. Building owners and architects look to meet the minimal safety standards of the building and fire codes. Generally, they feel that if the codes are met, adequate safety features will be provided, rather than minimal ones. As individuals, we tend to think that we will never get caught in a fire. We assume that buildings are safe and that, in the event of a fire, we can get to safety. Unfortunately, this is not true; this mind set could be life threatening.

Today's A/Es realize that fire safety and protection are an integral part of the design phase. A/Es have a responsibility to design fire-safe structures and they now consider how fires begin and spread when they design their buildings. Materials used during the constructions phase are carefully planned and discussed with the owner and support staff. A/Es incorporate fire safety measures into their design and use appropriate materials to design a fire-safe structure. Other A/Es and owners may not take fire safety into consideration; they will meet the minimum standards.

The security/FSD whose input regarding fire safety is requested by the owner at the inception of the design phase, has the advantage over the FSD who is consulted after building construction. Because NFPA standards are referenced by model building codes, the security/FSD can do very little to dispute the codes. However, his recommendations and advice to the A/E regarding security and fire safety systems are vital.

FIRE SAFETY TERMS

Operative terms, regarding fire-safe buildings, that are incorporated in an A/E's vocabulary include

- fire safety
- fire prevention
- fire growth hazard
- fire load
- fire suppression systems

Fire Safety and Prevention

In simple terms, the A/E's overall objective is to design a fire-safe building. A/Es incorporate fire prevention features into their designs that prevent fires from starting or spreading. Fire prevention can be achieved by using materials that are classified as *low-hazard materials*. Many residential and educational structures are classified as *low-hazard structures* because their rooms usually contain low fuel loads. Some mercantile and industrial buildings are classified as highly hazardous because they contain high fuel loads.

A/Es and FSDs wish to contain fires at their place of origin to prevent fire growth. The primary cause of fire growth is the combustion characteristics in a room. "The main factors that influence the likelihood and speed with which full room involvement occurs are: (1) fuel load (type of materials and their distribution), (2) interior finish of the room, (3) air supply, and (4) size, shape, and construction of the room."[35] These factors also affect the facility's fire load. The amount of combustible and noncombustible material used and stored in a building should be seriously considered during the initial design and interior design phases.

Fire Growth Hazard and Fire Load

Fire growth and *fire load* are terms that may be confusing. Fire growth is equivalent to the time it takes for a fire to spread. Fire load is equivalent to the amount of furnishings you have in a specific area at the time of the fire. Experts in the field of fire science also express fire load as the weight of combustible material per square foot of the fire area. Fire load is the direct cause of the fire growth.

If a fire occurs in a facility that manufactures clothing, imagine how fast that fire would spread. If your facility is a plant that manufactures steel, the likelihood of a fire spreading at a rapid rate is less than if a fire started at a clothing manufacturer. Fire duration is primarily based on the quantity of material available to burn. While an initial assumption is that fire growth will be less in the steel manufacturing facility, be aware that other types of materials may be stored in the steel manufacturing plant that may act as accelerants and cause other materials to burn as rapidly as the clothes at the clothing manufacturer's facility.

Fire Suppression

Fire suppression is the key to fire safety and fire prevention. By incorporating fire suppression systems into the structure during the design phase, the A/E is essentially saying, "I am aware a fire can occur at any given time and I want to reduce the possibilities of loss of life and property damage by incorporating a sound fire suppression system into the structure."

The most widely used technique of restricting the spread of fire is the automatic sprinkler system. The benefit of a sprinkler system is that it actually functions directly over a fire and is not affected by smoke or toxic gasses.

As heat is transferred by conduction, convection, and radiation, the most effective fire suppression system should stop the heat and fire from being transferred from one location to another.

Every A/E is confronted with the fact that a structure may collapse due to the use of inadequate materials or poor planning. Building codes address this possibility in their construction requirements. The correlation between fire severity and the vulnerability of the structure to collapse is of paramount consideration in all phases of construction. "Collapse can occur when the fire severity exceeds the fire endurance of the structural frame. However, this is comparatively rare. Structural collapse is more commonly associated with deficiencies in construction. These deficiencies are not evident under normal, everyday use of the building. They become a problem when the fire weakens supporting members, triggering a progressive collapse."[36]

The cost of sprinkler systems must be considered as well as their effectiveness. An effective sprinkler system is expensive and requires maintenance, but it is the most effective fire-fighting device.

BUILDING CONSTRUCTION

Architects, construction engineers, and insurance companies have spent a considerable amount of time and have used logical methods of structure design to oppose and combat the effects of fire. "It, therefore, also has become apparent that traditional methods of fire design using 'deemed to satisfy' clauses in Statutory Regulations or Design Codes of Practice are inadequate due mainly to their conservative and restrictive nature. Equally it has become recognized that the standard furnace test bears little or no relation to the actual structure behavior in a real fire."[37] It is necessary to propose idealistic methods of fire safety during the design phase and seek fire engineering input when building construction begins. Architects cannot calculate specific conditions into their designs. They can, however, calculate fire safety and fire prevention measures into their structures.

Over the past few years, fire engineering design through research has caused A/Es to specify areas of importance that are crucial to their designs. Fire and structural engineers study building designs and the materials used in constructing a building. Also, much is learned from a building that has been exposed to fire. The five crucial areas in fire engineering and research are

1. *Fire Protection*. In this section the basic need for Fire Protection and the modeling of fire response is dealt with.
2. *Material Behavior*. This section deals with modeling the behavior of basic construction materials at elevated temperatures and their use in calculation methods.
3. *Design Concepts*. This section deals with the philosophy behind fire design—namely an extension of limit state methods—including computer applications.
4. *Design Implementation*. This considers the different approaches required for design in each of the common structural materials.
5 *Post–Fire*. The last section considers what happens after the fire and the measures needed to reinstate the structure.[38]

The A/E who seeks the advice of a knowledgeable fire engineer gains a tremendous asset and the security/FSD attempting to understand the methods of design and application will begin to see that security and fire safety are taken seriously. Remember, the A/E is seeking to design a safe and secure structure, adhering to building and fire codes, while being cost-conscious at all times.

Structural engineers, A/Es, and building and fire officials have almost entirely relied on results of standard fire tests to evaluate the fire resistance of buildings and their components. In the United States, structural and fire engineers use the American Society of Testing and Materials (ASTM) E119 fire test. This test is applied to concrete structures. For this test, a specimen of a wall, column, beam, floor, or roof is subjected to a *standard fire*. Defined, a *standard fire* means a time-temperature correlation. During this fire test, load-bearing components must sustain loads that simulate the effects of the actual components. The *fire stamina* of these components is the duration of the fire test until the conclusion is reached. Structural engineers often refer to these end points as *stability, integrity,* and *insulation* (SII).

In the United States when concrete is fire tested, it is classified by aggregate type and unit weight. There are four types of structural concrete that engineers fire test:

1. *Siliceous aggregate concrete*—This concrete is composed of a normal weight aggregate of mainly silica or compounds other than calcium or magnesium carbonate.
2. *Carbonate aggregate concrete*—This concrete is made with aggregates of mainly calcium or magnesium carbonate.
3. *Sand lightweight concrete*—This concrete is produced with a combination of expanded clay, shale, slag, or slate, and natural sand.
4. *Lightweight aggregate concrete*—This concrete is made with aggregates of expanded clay, shale, slag, slate, or sintered fly ash. This concrete is generally lighter in weight compared to the sand lightweight concrete.

Structural engineers incorporate fire protection measures in association with fire testing steel, steel work, wood, timber, and, of course, concrete. Through fire-testing methods, structural engineers carry out strength behavior tests on these properties. Stress-related tests are also carried out regarding how these materials react during a fire. When materials are tested, results are analyzed to explain how materials react under various types of strain. The engineer seeks to expand fire-resistant periods during a fire. If certain materials enhance the fire-safety properties of a building and reduce the chances of a fire spreading, they are beneficial to all concerned.

Research has revealed that at elevated temperatures, the strength of materials does not depend solely on the temperature. At high temperatures *creep* develops that, depending on the temperature of the steel, can cause structure failure sooner than one could expect, judging by the strength-temperature relationship.

Fire resistance classifications are based on fire tests in accordance with the specifications of ASTM E119. These tests are supervised and administered by nationally recognized agencies. The classifications these tests receive are usually accepted by the building code authorities and the fire insurance rating bureaus. Fire classifications are the result of comprehensive examinations conducted on assemblies constituting distinctive materials that are put together in a specified manner. Agencies that certify tests include

- Building Materials & Structures, National Bureau of Standards (BMS)
- Factory Mutual Research Corporation (FM)
- George E. Troxell, PE, Consulting Engineer (GET)
- National Bureau of Standards (NBS)
- Ohio State University (OSU)
- Portland Cement Association (PCA)
- Standard Fire Test, Fire Prevention Institute (SFT)
- University of California (UC)
- Underwriters Laboratories Incorporated (UL)

In addition to evaluating the testing of fire-safe materials, the two primary considerations of these testing agencies are the proficiency of the structural support to avert collapse, and the proficiency of the barricades to avert ignition that causes heat and flames to spread to adjacent areas.

Fire Resistance Ratings

When A/Es begin to design their building, they must bear in mind that their structures must meet the standards of the building and fire codes. All fire-resistive assemblies should be fire stopped, or designed to prevent the spread of fire, in concurrence with the current building code requirements.

Fire resistance is the length of time the component being tested withstood the fire test without failure. "While the actual time is recorded to the nearest integral minute, fire resistance ratings are given in standard intervals. The usual fire resistance ratings for all types of members, structural assemblies, doors, and windows are 15 min, 30 min, 45 min, 1 hr, 1½ hr, 2 hr, 3 hr, and 4 hr. Therefore, a 1 hr rating indicates that the assembly withstood the standard test for one hour or longer. A 2 hr rating indicates that the assembly withstood the standard test for longer than 2 hrs without failure by any one of the failure criteria listed in the fire test protocol."[39]

The A/E who installs smokeproof enclosures and pressurized stairs must be sure compliance is met prior to any system being utilized. Prior to mechanical equipment being accepted at a facility, the A/E must be sure the systems were tested and confirm the equipment is operating in compliance with the various codes. Once these systems are installed and operational, the security/FSD is responsible for testing the systems semiannually and recording all results.

The security/FSD should know the terms used by A/Es and their support staffs regarding the architectural design of the structure. *Fire stopping* is often associated with the design of a structure. It is crucial for the A/E to design, evaluate, and select fire-stopping materials for a structure." The fire-stopping requirements for partition walls usually require that the wall cavities contain horizontally attached members that block the vertical space between framing members. This eliminates the chimney draft effect within stud spaces that could contribute to the spread of fire from one protected area to another. All fire resistive assemblies should be fire-stopped in accordance with the prevailing building code requirements."[40]

After a facility sustains an actual fire, the structure is inspected. Engineers will examine fire severity, structural design, material properties and occupancy. Much information can be obtained from a building after a fire. For example, in 1976, the McCormick Place fire occurred; at the time, it was the nation's largest exhibition center. The structure was divided into three levels. The lower and immediate levels were constructed of reinforced concrete. The upper level and roof were constructed of structural steel. Sometime in the early morning hours, the day a semiannual exhibit was to take place, a fire began in an exhibition booth on the upper level. The fire spread and caused much property damage. After the fire, the investigation revealed that ineffective fire-fighting techniques by untrained personnel; a time delay in sounding the first alarm; lack of sprinkler protection; inoperative private water supply, pumps, and controls; inoperative water supply mains and hydrants; and combustible materials contributed to the rapid growth of this fire. The fire was not under control until *7 hours* after it began! That is far too long a time period.

The structure was severely damaged; practically the entire roof of the structure collapsed. The reinforced concrete construction on the top level was not seriously destroyed. Documentation indicates that the reason the roof collapsed was due to the unprotected steel being unable to endure the intense temperatures. Sometime later, McCormick Place was rebuilt into a larger exhibition center. Structural and fire protection engineers made many changes. Steel members were coated with fire-protected material. A heavier metal was incorporated into the new steel frame and operational sprinkler systems were installed.

Exits

In buildings under construction, adequate means of escape must be maintained at all times. Escape routes and portals consist of doors, walkways, stairs, ramps, fire escapes, and ladders. Every exit, exit access, or exit discharge must be continuously maintained and devoid of obstacles. In the event of a fire condition, exits must be accessible to evacuees, the local fire brigade, and the fire department.

In the NFPA 101 *Code for Safety to Life from Fire in Buildings and Structures,* 1988 edition, egress and travel distance is widely discussed. A/Es and security/FSDs are, at times, perplexed by these requirements. With regard to exits, this rule shall always apply: "The travel distance to an exit shall be measured on the floor or other walking surface along the center line of the natural path of travel starting 1 ft (30.5 cm) from the most remote point, curving around any corners or obstructions with a 1-ft doorway or other point at which the exit begins. Where measurement includes stairs, the measurement shall be taken in the plane of the tread nosing."[41] The exception to this rule applies to correctional facilities. Therefore, the Life Safety Code sets precedents regarding measurement of travel distance to exits.

In regards to travel distance limitations, "travel distance to at least one exit shall not exceed 200 ft (60 m) in buildings not sprinklered or exceed 250 ft (76 m) in buildings protected throughout by an approved supervised sprinkler system in accordance with section 7-7."[42] The two exceptions to the distance limitations rule apply to NFPA 101, chapters 8 through 20, and travel distance regarding facilities storing hazardous materials. It is appropriate to say that specific rules apply to the various codes established by the NFPA. Become familiar with the NFPA codes and learn how they effect your facility.

The FSD must incorporate variables into her thought process regarding egress. These variables include:

1. a survey of the facility's entire fire protection scheme
2. an appraisal of the population's characteristics
3. potential dangers

Egress is an indispensable segment of the design phase and fire operation plan. Many studies have been conducted of exits after a fire, including the reactions of persons who survived the fire. The reactions of individuals vary greatly.

As the security/FSD, visualize your facility and observe how many exits lead to the street or to other areas of the facility. According to the NFPA 101 code, "All exists shall terminate directly at a public way or at an exit discharge. Yards, courts, open spaces, or other portions of the exit discharge shall be of required width and size to provide all occupants with a safe access to a public way."[43] If any exits in your facility, at present, do not comply with this regulation, seek the advice of an A/E or fire protection engineer. (Remember the X-factor.) In the years to come, more studies will deal with the psychological factors regarding egress and fire safety.

Means of egress are vital in researching human behavior during a fire condition. Human behavior and the individual's ability to think rationally during a fire evacuation are crucial. Behavior during a fire and the interactions between individuals is paramount. According to an article by Jake Pauls in the *Fire Technology Journal,*

What is most important—and here we have failed badly in the egress field—is to continue to examine critically the methods and data adopted in design and regulatory practices. While we spend a billion or so dollars annually to construct means of egress (not to mention the costs of unrentable space) we invest almost nothing in egress research.[44]

On May 28, 1977, there was a tremendous fire at the Beverly Hills Supper Club in Kentucky. At the time of the fire, the building was occupied by an undetermined number of people, but the number was known to be in excess of 1,600 patrons and approximately 182 employees. One hundred sixty-four persons died in that fire. The major cause of death was smoke inhalation and acute carbon monoxide poisoning.

After the fire the investigation began, means of egress were investigated by a panel appointed by the Governor of Kentucky. The panel found that "the magnitude of the loss of the life and injury in the Cabaret Room is well documented and, in the opinion of the Investigating Team, attributable in large part to inadequate means of egress for occupants of that area."[45] It was the further opinion of the investigating team that the Beverly Hills Supper Club was constructed with insufficient exits.

At present, there are no current egress requirements pertaining to evacuating a mass of occupants from a specific area. Current egress requirements were developed decades before relevant research was completed. Crowd movement in a fire emergency causes confusion and panic. Without guidance, crowds flow to where they believe unobstructed passage exists.

When emotion takes over, human behavior cannot be predicted. Studies have shown that when egress is denied to persons fleeing a fire condition, behavior changes. "The model of human behavior in a fire emergency consists of three stages: detection of cues, definition of the situation, and coping behavior. In this model, the individual is depicted basically as a reactor to the environment; however, instances are included in the fire coping stage in which the individual impacts on the environment."[46] Once egress is denied, individuals begin seeking alternate areas of refuge as a group. This is also known as *convergence clusters*.

Occupancy Load

Occupancy load is based on the number of persons present in a building at any given time for whom egress must be available. This figure is calculated by dividing the gross area of the facility (or the net area of a specific segment of the facility) by the area in square feet (or square meters) that is predetermined for each individual. The amount of predetermined floor area varies with the structure.

"In some situations the actual number of people in a building can be determined at the design stage, in which case this number should be used in the design of exits. A typical example of this is an assembly occupancy in which fixed seating is installed. Counting the number of seats provided would obviously give a more accurate figure than multiplying a square foot (or square meter) per person figure by the net floor area."[47]

BUILDING CONSTRUCTION CLASSIFICATION

As the security/FSD, you will be asked numerous questions pertaining to the classification of your building. It is important to understand what is meant by *building classification*. If the building is already constructed, ask to see the certificate of occupancy. The *C of O* (as this certificate is sometimes referred) indicates that all building-related criteria and other germane codes have been met, and the facility is safe and habitable.

"A well established means of codifying fire protection and fire safety requirements for buildings is to classify them by types of construction based upon the materials used for the structural elements, and the degree of fire resistance afforded by each element."[48] Learn about the construction of your facility and the materials used in its construction. This knowledge will aid and benefit you in developing your fire safety plan and fire operation plan. The following sections describe the various types of construction classifications.

Type I Construction (Fire Resistive)

"Type I construction, fire resistive, is construction in which the structural members are noncombustible and are fire-protected as specified. In Type I construction only noncombustible materials are permitted for the structural elements of the building. This is an accepted regulation that appears in practically every modern building code. Some codes have attempted to regulate combustible materials by using a definition of noncombustible materials which includes two or three alternatives that allow for the acceptance of materials having relatively low fuel content and surface burning characteristics."

Type II Construction (Noncombustible)

"Type II construction (noncombustible) is a construction type in which the structural elements are entirely of noncombustible or limited combustible materials permitted by the code being applied and either protected to have some degree of fire resistance, either 2 hr [Type II (222)], 1 hr [Type II (111)], or completely unprotected except for exterior walls, Type II (000) construction. Height limits, however, are commonly prescribed for this type of construction. The noncombustible feature is valuable because it prevents fires from spreading through concealed spaces or involving the structure itself.''

Type III Construction (Exterior Protected Combustible)

''Type III construction (exterior protected noncombustible) is a construction type in which all or part of the interior structural elements may be of combustible materials or any other material permitted by the particular building code being applied. The exterior walls are required to be of noncombustible or limited noncombustible materials acceptable to the code and have a degree of fire resistance depending upon the horizontal separation and the fire load.''

Type III (200) construction has no protection for the floors or structural elements. Whether or not fire protection is provided, it is essential that all concealed spaces be properly fire stopped in buildings of combustible construction.

Type IV Construction (Heavy Timber)

''Type IV construction (heavy timber) is a construction type in which structural members, i.e., columns, beams, arches, floors, and roofs, are basically unprotected wood (solid or laminated) with large cross-sectional areas. No concealed spaces are permitted in the floors and roofs or other structural members with minor exceptions.''[49]

In Type IV constructions, exterior and interior walls can be constructed of noncombustible or limited combustible material. However, check with the building codes to ensure that the specifications are met. In the event of a fire condition, it is known that timber withstands collapse longer than a customary wood-framed structure.

Type V Construction (Wood Frame)

''Type V construction (wood frame) is a type of construction in which the structural members are entirely of wood or other material permitted by the code being applied. Depending upon the exterior horizontal separation, the exterior walls may or may not be required to be fire resistive.''

Type V construction is probably more vulnerable to fire, both internally and externally, than any other building type. Fire-stopping in exterior and interior walls at ceiling and floor levels, in furred spaces, and other concealed spaces can retard the spread of fire and hot gases in these vulnerable areas.

Mixed Types of Construction

"When two or more types of construction are used in the same building, it is generally recognized that the requirements for occupancy or height and area would apply for the least fire resistive type of construction, However, in cases where each building type is separated by adequate fire walls or area separation, each portion may be considered as a separate building.

"Another general limitation included in some codes prohibits construction types of lesser fire resistance to support construction types having higher required fire resistance. In the event of a fire, the risks of a major structural collapse are generally too great to permit this type of design."[50]

SUMMARY

The A/Es responsibilities include hiring competent support staffs to assist them during the design phase. A/Es will consult with mechanical, electrical, plumbing, and fire engineers to design and ensure the construction of a safe and secure structure. They will also determine the structure's construction classification based on the building's occupancy. The criticality and vulnerability of the structure depends on the type of structure being designed by the A/E.

Thomas J. Whittle, PE, discusses a six-phase design sequence for the security/FSD to refer to when a building is under construction:

1. study and report phase
2. preliminary design phase
3. final design phase
4. bidding and negotiation phase
5. construction phase
6. operational phase

These phases allow A/Es to incorporate security and fire safety measures from the building's inception. A/Es and their support staffs should ask for the security/FSD's input regarding security and fire safety during these phases.

Much has been learned through research and design of a building, especially through testing building materials. Other valuable information has been obtained from actual building fires. Research and design in five critical areas of study has enhanced the A/Es and fire engineers jobs—fire protection, material behavior, design concepts, design implementation, and post-fire strategies. The standard fire tests, which evaluate buildings and their components, are being evaluated and studied continually.

Most buildings are designed of concrete and steel. The classification of concrete and how it reacts to a fire condition is important knowledge for the A/E. Fire resistance classifications are based on fire tests in accordance with the specifications of the American Society of Testing & Materials. All fire tests are supervised and administered by nationally recognized agencies. It is important for security/FSDs to know how the fire resistance ratings are issued and how they effect their facilities.

The definitions issued by the NFPA regarding exits and travel distance in occupancies are based on life safety codes; structures must comply with these codes. Egress and convergence clusters are areas the security/FSD must consider. However, this can hardly be incorporated into a fire safety plan.

The classification of buildings and the certificate of occupancy, or C of O, is clearly defined. In order to be issued a C of O, all criteria applicable to the building and other germane codes must be met and the facility must be safe and habitable. During building and fire inspections, the FSD will be issued a summons if these criteria are not met. The six types of construction classifications are fire resistive, noncombustible, exterior protected combustible, heavy timber, wood frame, and combined types.

The security/FSD should be aware of the obstacles that an A/E encounters during the design of the structure.

Many of these same obstacles exist if your facility is under renovation. The security/FSD can best cope with these problems by supplying the A/E with well-researched information and recommendations regarding fire systems and their costs.

8

Liability and Fire Prevention

Security operations and loss prevention programs work to ensure the protection of life and property, including patrolling buildings of almost any type and classification. Your security strategy should not only include protecting the property from criminal activity, but should also include protecting the owner and the property from negligent activities (such as fire) that could result in property and monetary loss. It is imperative that the security/FSD be aware of all local ordinances and amendments concerning the elimination of fire hazards and the disposal of hazardous waste materials.

Let us examine the three types of liability with which the security/FSD, administration, and their staffs will be involved:

- criminal liability
- strict liability
- vicarious liability

CRIMINAL LIABILITY

"As with tort law, criminal law is concerned with the enforcement of legal duties to act, not moral duties to act. A legal duty may arise as the result of a contract (e.g., a lifeguard obligates himself to attempt to rescue the drowning swimmer); by virtue of a special relationship (e.g., a parent has a duty to provide for the welfare of his child); by statute (e.g., a hospital has a duty to provide emergency care); or by common law (e.g., where one places another in peril, there is a duty to make reasonable attempts to assist the one endangered)."[51] Criminal liability can also be defined as an obligation to do or refrain from doing something illegal.

Elements of criminal liability are the voluntary act or omission to act, *actus reus*; the mental state of the individual at the time of the act, *mens rea*; and the cause of the crime. Criminal liability must be outlined by security/

FSDs when they establish their fire safety and fire operation plans. Corporations and establishments, as well as individuals, can be found guilty of criminal liability.

State of Connecticut v. Gordon L. White

Criminal liability is the issue in the *State of Connecticut v. Gordon L. White*, July 14, 1987. Gordon L. White is charged with negligent homicide for causing the death of three occupants and with failing to provide smoke detectors in his residential building, which he owned and was occupied by three tenants.

On Christmas Day, 1982, in New Britain, Connecticut, a resident of a first-floor apartment was woken by the smell of smoke. He awoke his family members and escorted them to safety outside. He also notified the fire department, which arrived almost immediately to the building. The firemen fought the fire and, in the process, found the bodies of Mary Ann Jones and her two young children in a second-floor apartment. It was determined that the cause of death of all three occupants was due to asphyxiation from smoke inhalation. It was also determined that the fire originated in a second-floor apartment and was caused by an electrical overload in a wall outlet. It was estimated that the overload smoldered for approximately 3 hours before igniting. There were no smoke detectors installed anywhere in the building, nor in any of the apartments.

At the trial, evidence was presented showing that the existence of smoke detectors in the building may have prevented the deaths of the three victims. To make matters worse, Gordon L. White owned another property in East Hartford, Connecticut, and was warned by the fire marshal 6 months prior to the New Britain fire to install smoke detectors in that residence. Gordon L. White complied with the fire marshal's request and installed the smoke detectors in the East Hartford complex.

The defendant's motion for acquittal was denied by the trial judge. White was convicted on six counts—three counts of negligent homicide and three counts of failing to provide smoke detectors. He was sentenced to 1 year in jail, which was suspended, and placed on 3 years probation. He was also fined $4500. White appealed his conviction and sentence.

White's argument was rejected by the appellate court for his appeal that the fire safety code and the statute authorizing the establishment of the code were unconstitutionally vague. The court found that White had been "fairly warned that he had a duty to install smoke detectors in his New Britain Building." However, the appellate court did, in fact, agree with White's argument that the fire marshal's fire safety code regulations exceeded the scope of authority conferred by the enabling statute authorizing the marshal to promulgate the fire safety code. Prior to reaching its conclusions, the appellate court reviewed the state fire safety code provisions and applicable statutes. The legislative history behind the codes were researched as well.

In their research it was found that the court first reviewed the 1947 Connecticut law, which provided

> The state fire marshal shall establish a fire safety code and at any time may amend the same. The regulations in said code shall provide for reasonable safety from fire, smoke and panic therefrom, in all buildings except private dwellings occupied by one or two families, and upon all premises except those used for manufacturing. The purpose of the statute was to give the fire marshal the ability to enact reasonable minimum requirements for safety in new and existing buildings.[52]

This act remained unchanged for 30 years. However, in 1976, the legislature amended it by requiring that the regulations issued by the fire marshal

> shall include provision for smoke detection systems in residential buildings designed to be occupied by two or more families for which a building permit is issued on or after October 1, 1976.

The fire marshal then amended the fire safety code to require that smoke detectors be installed in

> each guest room, suite, or sleeping area of hotels, motels, lodging or rooming houses, and dormitories and in each dwelling unit within apartment houses and one- and two-family dwellings . . .[54]

With the research completed, the appellate court concluded that the legislative intent regarding the 1976 act was not clear. The statute directed that the fire code should provide regulations for the installation of smoke detectors in buildings for two or more families, providing a buiding permit had been issued on or after October 1, 1976.

The appellate court felt that the fire code exceeded the scope of its statutory authority by attempting to have smoke detectors in all buildings occupied by two or more families. The court concluded that White could not be legally convicted of violating the fire safety code requiring the installation of smoke detectors without proof that a building permit had been issued for the New Britain complex on or after October 1, 1976. Since the state could not prove this evidence at trial, the appellate court ruled that White's convictions would be reversed and set aside.

"The appellate court also reversed White's convictions for the three counts of criminally negligent homicide. In order to be found guilty of criminally negligent homicide, the defendant must have caused death while failing to perceive a substantial risk that the death will occur. The state would have to prove three elements:

(a)   that there was a duty or obligation on the defendant required by law, to conform to a certain standard of behavior;

(b)   that the defendant on trial had grossly deviated from such standard of behavior;

(c)   his deviation was the direct cause of the resulting death."[51]

The appellate court determined that White was not obligated under the fire safety code to install smoke detectors. Therefore, he could not be convicted of criminally negligent homicide. The appellate court reversed the defendant's convictions on all six counts.

To sustain a criminal conviction for negligent homicide, the state must prove beyond the shadow of a doubt that the defendant had deviated from standard conduct with which he was required to meet, which caused the death. This case is a criminal prosecution brought against White by the state of Connecticut. If, instead, a civil suit had been brought against White, the plaintiffs would have been seeking monetary damages.

There are four areas of culpability considered by the legal system when deciding cases—intentional, knowing, reckless, and criminal negligence. These areas are important to understand from an individual and corporate standpoint. If individuals or corporations act *intentionally* regarding a statute defining an offense, they do so when their conscious objective or motive causes loss of life or property damage. The court will deem this act of culpability *intentional*. If individuals or corporations are aware that their conduct violates the statute, the courts will deem that if the individual or corporations knew of the malfeasance prior to the offense, the courts will rule the violators acted *knowingly* against the statutes. If individuals or corporations act *recklessly* regarding a statute defining an offense, they are aware of and consciously disregard the possibility of a potential hazardous situation. The means used to measure the potential risk is how a *reasonable* person or corporation would act regarding a potential hazardous situation. If individuals or corporations act with *criminal negligence* regarding a statute defining an offense, they fail to recognize a substantial and unjustifiable risk of a given situation. The risk must be of such nature and degree that a reasonable individual would foresee the potential hazard and act to correct it.

In regard to criminal liability of corporations, it must be thoroughly understood that the courts view corporations in a different light than individuals. The terms used when discussing corporate liability include *agent* and *high managerial agent*.

An agent is any director, officer, employee, or any other individual who is empowered to act on behalf of the corporation. A high managerial agent is an individual with authority, such as an officer who supervises, manages, or directs employees and formulates corporate policy. Security and FSDs can be considered high managerial agents.

Corporations can be considered guilty of an offense if they fail to act according to a statute. If the courts find that a board of directors or a high managerial agent acted in a manner that solicited, authorized, requested, commanded, or recklessly tolerated offensive behavior, then they can be found guilty of criminal behavior. As long as an individual is employed by the corporation and the employee acts on behalf of the corporation, the corporation can be criminally liable for his actions.

It is beneficial, then, to have the security/FSD trained to prevent and deter criminal activity or fire-related activities from occurring in the workplace. Landlords or building managers are agents of a specific property. It is their responsibility to maintain and safeguard their tenants and guests. Their actions or lack of actions can be the cause of criminal or civil liability.

Lasheryl Williams v. Jerry Johns

An example of criminal negligence involving a tenant and a landlord was decided by the Michigan Court of Appeals in January 1987. The case, *Lasheryl Williams v. Jerry Johns,* is one in which the defendant, Jerry Johns, owned an apartment building in Detroit, Michigan. Williams was a tenant in the apartment building owned by Johns. Late one evening, the building caught fire and Williams was forced to jump out of her third floor apartment window to escape the flames. She was seriously injured and hospitalized as a result of her fall. She initiated a lawsuit against Johns and claimed that Johns was responsible for her injuries since his negligence created an unsafe and hazardous condition. She further claimed that Johns was negligent, because he stored combustible waste material in the building. She also alleged that Johns violated two city ordinances regarding the storage of combustible waste material by storing them inside the building.

The Detroit arson investigation team concluded that the fire originated in trash stored at the bottom of a stairwell near the back exit of the building. According to the arson investigation team, the fire began when an unknown individual deliberately set fire to the waste material at the bottom of the stairwell. This caused the fire to spread rapidly throughout the interior of the building. The stairwell inside the building acted like a chimney and quickly vented the flames and smoke to the upper floors. The interior stairwell could not be used as a means of egress.

Johns moved for dismissal of the case. His attorney argued that even if Williams' allegations were correct, negligence was not an issue, since the fire was the result of arson and, therefore, Johns could not be considered responsible for Williams' injuries.

Williams responded by arguing that Johns' actions of allowing the combustible material to be stored in the building stairwell violated two of the city ordinances. This constituted negligence on his behalf. Williams' argument was that regardless of who set the fire, Johns contributed to it by storing the combustible material in the building.

The trial judge concurred with Johns and granted his motion for summary judgment and dismissed the case. *Summary judgment* is a preverdict judgment rendered by the court in response to a motion by a plaintiff or defendant. The plaintiff or defendant usually claim that there is absence of factual dispute on one or more issues, summary judgment eliminates the need to send those issues to a jury.

Williams appealed. The Michigan Court of Appeals first noted the ordinances of the city of Detroit. The ordinance pertaining to the fire prevention code defines a hazard as

> Any situation, process, material or condition which on the basis of applicable data, may cause a fire or explosion or provide a ready fuel supply to augment the spread or intensity of the fire or explosion and which poses a threat to life or property of others.[56]

The appellate court further noted that a landlord "owes a duty to his tenants to protect them from unreasonable risks of harm resulting from foreseeable activities occurring within the common areas of the landlord's premises"[57] and determined that the relevant question was whether the landlord's storage of combustible material in the stairwell constituted negligence on his part.[53] The court found that this issue was a question of fact and it should have been decided by a jury, not by a judge. The court ruled that the trial judge erred in making a motion for summary judgment. The appellate court further ruled that the trial judge also erred by determining that arson precluded the plaintiff from proving that the landlord's negligent conduct, if any, had caused injuries to the tenant.

The appellate court in this regard noted that it is not necessary for a plaintiff to prove that the defendant's negligence was the only causal factor resulting in harm to the plaintiff; it is sufficient to establish liability if the plaintiff can prove that the defendant's negligence was a substantial factor in causing the harm incurred. The fact that there was an intervening cause, such as the arsonist using the waste material to set the fire, does not constitute an absolute bar to the landlord's liability if the intervening act is foreseeable even though the act itself may be negligent or even criminal.[58] The appellate court concluded by stating that a jury should have the opportunity to resolve these issues and returned the case to the lower court for retrial.

As a security/FSD, you or a building manager could be placed in a similar situation. It is imperative that you understand the local ordinances regarding eliminating fire hazards or disposing of hazardous materials. Your inaction could be cause for a negligence suit or even criminal prosecution. This case is a classic example of how the courts will determine whether the owners or building managers have or have not carried out their duties to residents, visitors, and employees in protecting them from foreseeable dangers.

To help you avoid this situation, use a *monthly fire protection inspection checklist.* (See Figure 8–1) This checklist can be used to advise the security/FSD of possible hazards throughout the facility. Unlike Figure 2–1, the Fire Safety Audit Sheet, this checklist details specific equipment to be used in pre-

Date of Inspection: _____

Inspector's Name: _____

Address of Facility: _____

Equipment	Location	Remarks
Sprinkler system:		
How many sprinkler heads?	_____	_____
Shut-off valve operational?	_____	_____
Outside stem and yolk valve:	_____	_____
Flow alarms:	_____	_____
Dry pipe system:		
Is pressure adequate?	_____	_____
Low-pressure alarm system:	_____	_____
Storage tanks:	_____	_____
Is pressure adequate?	_____	_____
Low-pressure alarm system:	_____	_____
Standpipe and hose equipment:	_____	_____
Are hoses operational?	_____	_____
Is hose attached to the standpipe with a nozzle?	_____	_____
Fire extinguishers:	_____	_____
How many?	_____	_____
Are they operational?	_____	_____
Do they have their inspection tags?	_____	_____
List types (A,B,C,D):	_____	_____
Fire hazards:		
Paint rooms		
Furniture storage rooms	_____	_____
Cooking locations	_____	_____
Other	_____	_____
Inspector's Signature		

FSD Signature _____ Date: _____

Figure 8–1 Monthly fire protection inspection checklist.

venting fires and allows the inspector to make remarks regarding the inspection. The security/FSD should review this information on a monthly basis and then review the findings with the building's administration.

As the security/FSD, you are always seeking ways to avoid premises liability. If your department and the housekeeping or operations department incorporate good housekeeping and fire prevention measures into their everyday operations, the risk of liability is substantially reduced.

In many instances, when lawsuits for criminal or negligent liability are brought against a corporation or individual, everyone and anything can be sued. As the security/FSD for your facility, you can be sued, along with your corporation, for negligence. Criminal liability is most frequently imposed on private security officers in cases involving the use of force against others. The officer can be charged with assault, battery, manslaughter, and murder. Be sure your security department is aware of criminal liability and its impact.

STRICT LIABILITY

Strict liability, simply defined in tort and criminal law, is liability without fault. *Tort law* is a private or civil wrong resulting from a breach of legal duty that exists by virtue of what individuals expect regarding interpersonal behavior, rather than by behavior or other personal relationships. The integral component of a tort is a lawful obligation owed by a defendant to a plaintiff.

Certain criteria signify criminal liability, strict liability, and mental culpability. The criteria for criminal liability are the display of conduct by an individual that consists of a free-will act or the exclusion to perform an act that the individual is physically capable of performing.

FORESEEABILITY

As the security/FSD, *foreseeability* is a term with which you should become acquainted. Strictly defined, foreseeability means having prior knowledge that certain events could occur. Many times, a corporation's liability may arise from the condition of the property or the building. Most courts will agree that the duty to randomly conduct reasonable investigations of the property in order to seek and observe dangers must be performed by property owners or corporations. The hazards, once discovered, should be mitigated or eliminated. Let's take an example in which your facility is storing paint cans on-site, in a basement room. The room is not equipped with a sprinkler system, it is secured with a special key and there are no *No Smoking* signs in the immediate area. As the security/FSD, you have a tremendous problem. Why? There is a *foreseeable* problem that is a fire hazard. As the security/FSD, you are responsible for remedying the situation, as this is a no-win situation for your corporation.

Correct this situation by having the paint cans moved to an area that is protected with a fire suppression system or that is isolated from the remainder of the facility. Be sure that the door is accessible 24 hours a day and place signs in the room and in the immediate area that state *No Smoking Hazardous Materials Being Stored.*

In the event that your fire and burglar alarm system malfunctions, it is your responsibility to correct the situation and have the system operational within a reasonable amount of time (i.e., 24 hours). If your security and fire systems are tied into a central station and this system malfunctions, it is your responsibility to notify the central station that the system has malfunctioned and then notify your alarm company to promptly come to your facility and correct the malfunction. All of this information should be logged in your security log book and an incident report should be forwarded to administration explaining the situation. As an added precaution, you may want to call the local fire company and explain that your fire alarm system is inoperable. In this way, you are taking the necessary precautions to correct a hazardous situation.

Haynes and North River Insurance v. Aetna Casualty and Miller's Mutual

In March 1982, The Louisiana Court of Appeals decided the case of George D. Haynes, d/b/a Geo-Je's Apparel, and North River Insurance Company versus Aetna Casualty & Surety Company and Miller's Mutual of Texas. The premise of the suit was the disconnection of the fire/burglar alarm system by unknown individuals. The case was based on the question, Was it sufficient to hold the store owner liable when the fire spreads to another's property?

Geo-Je's Apparel was damaged by fire when an unknown third party disconnected Lee Allen's Fashion for Men's alarm system and then deliberately set fire to the store. The fire spread, causing damage to some stores attached to Lee Allen's Fashion for Men. Many law suits resulted from the fire; only the dismissal of Geo-Je's claim against Allen's insurer was appealed.

The burden of proof was on Geo-Je's Apparel. They had to prove that the insured was responsible for the cause of the fire. The only testimony introduced at the trial, to implicate the proprietor of Allen's with starting the fire, was gathered from the journals of the alarm company that installed the alarm. The arson investigator who reviewed these records testified that at 6:15 P.M., on the night of the fire, approximately 10 minutes after Allen's owner and employees departed from the store, an intruder disconnected the fire/burglar alarm system. The intruder had knowledge of the alarm system. After the intruder set fire to the store, he reset the alarm. At 6:23 P.M., the alarm company reported that it received an active alarm from Allen's store.

"This evidence, based mainly on hearsay, was insufficient to identify the arsonist and impose liability on the insurer. One of Geo-Je's arguments was that Allen was strictly liable for the damages caused by the spreading of the fire, no matter what the cause. This position was based on a Louisiana statute which covers situations where the owner or lessee of property is carrying on some activity on his property which causes damages to his neighbor."[59]

The Louisiana Court of Appeals rejected this argument. They cited another case in which a mysterious fire spread to a neighboring cane field:

> The (Louisiana statute) has no application to the circumstances of this case. To hold otherwise would cast liability on every land owner on whose property a fire is commenced and spread to neighboring lands even

though the cause of origin of the fire was due to the act of trespassers, vandals, lightning, or unknown sources not attributable to any conduct or activity on the part of the landowner.[60]

In this case, the Louisiana Court of Appeals affirmed the dismissal of the action against Allen's insurer. It is important to note that in this case, strict liability could not be imposed because there was no foreseeable knowledge that an intruder would break into the store, then set it on fire.

A major problem with foreseeability is that it is too vague to legally define. Courts can be arbitrary in their definition of the word. As the security/FSD, you will have difficulty explaining to your administration exactly what the courts view as *reasonable precautions*.

If your facility has installed or is in the process of installing fire detection systems, it is imperative that the manufacturers of these systems take the necessary precautions to make sure these systems are operational and useful in your facility. The primary purpose of a fire detection system is to note irregular environmental conditions, such as the appearance of smoke, light fluctuations, temperature elevation, or radiation. These detectors function on resources that consist of thermal expansion, thermoelectric sensitivity, thermoconductivity, and photosensitivity. As the FSD, you must formulate and use a checklist to ensure that all systems are operational and functional. Should any part of your fire detection system fail, and you were aware that the system was inoperative, you would be held liable and could be charged with negligence.

The security/FSD who has a solid background in law enforcement usually has little or no knowledge of fire safety. The only alarm system with which he is familiar is a burglar alarm. The burglar alarm has proven effective in deterring crimes and, many times, in apprehending criminals. The problem with burglar alarms is the still-to-be-resolved false alarm rate. "From 90 to 98 percent of all alarms transmitted are said to be false. This high percentage can be basically attributed to three factors: (1) user error or negligence, (2) poor installation or servicing, and (3) faulty equipment."[61] However, as the FSD, you cannot assume that your fire alarm system has malfunctioned. Each and every time the system is activated, it must be responded to. The one time to which it is not responded and a fire occurs, when you have previous knowledge that the system is inoperative, you will be liable.

An FSD at a hotel in Chicago, Illinois, advised me that when his staff receives an active alarm, they follow their fire operation plan and respond to the problem area. However, when the system malfunctions and refuses to shut itself down, it can cause concern. The FSD informed me that until the system is operational, he assigns one of his employees to fire patrol. He then notifies the alarm company of the problem. Until the alarm company responds to correct the system malfunction, his employee will conduct fire patrols. From a liability standpoint, this is an excellent idea. What more could an insurance company ask regarding your reaction to an immediate problem?

CONTRACTUAL LIABILITY

As the security/FSD, you rarely become involved with the insurance company. The corporation's administrative office or the business agent will select the best possible insurance coverage for the organization. Yet, it is important for you to be acquainted with certain terminology. An important aspect of a liability provision with which you should be familiar is *contractual liability.*

Your company as well as yourself can, by means of a written or oral contract, be held liable for the negligent acts of another. For example, if a contract guard company provides your security and fire staff and a fire occurs due to their negligence, you and your organization can be held liable.

Legally, liability for the negligent behavior of another person is incurred or implied through a contractual agreement. Often, provisions are contained in the contractual liability agreement that state that one party agrees to hold the other party blameless for all claims for injuries arising out of the performance of the work. This is also referred to as a *hold harmless clause.* If this provision is put into a contract the person assuming the liability (the *indemnitor*) agrees to pay monetary damages stemming from the work that may be entered against the other party (the *indemnitee*).

Ask about the various types of contractual liability that may affect your department. Many times, contractual liability is automatically covered under premises and operations insurance coverage. However, other insurance contracts must be specifically insured for contractual liability.

VICARIOUS LIABILITY

Vicarious liability places the imputation of liability on one person for the actions of another. As an example, in tort law, if an employee, while in the scope of her employment duties, drives her vehicle and injures a pedestrian, the employer will be held vicariously liable.

In some jurisdictions, various criminal laws may apply. As an example, a bartender employed by a nightclub sells liquor to a minor. The nightclub is responsible for the actions of the bartender. In some jurisdictions, this is also known as *imputed liability.*

FSDs as well as their security and fire safety officers who engage in tortious conduct are individually liable for the damages caused by their actions. FSDs and their staffs are usually employed by an organization or corporation that is more solvent than themselves. Therefore, a plaintiff will sue the employer as well as the employee and attempt to establish liability under the doctrine of respondent superior.

The doctrine of respondent superior is based on axioms of agency law. *Agency law* refers to a circumstance in which two parties concur that one is to act in the interest of the other according to the latter's instructions. An agency originates where one individual, the *principal,* has a perogative to supervise the action of the other individual, the *agent*. As an FSD, you are the principal and your officers are agents. Therefore, the organization, you, and your staff can be sued if it is deemed that you acted negligently.

The doctrine of respondent superior also applies to a circumstance in which a master, the *employer*, is liable for the torts of the servant, an *employee*. Employers are legally accountable for employees' deeds when these deeds occur under the scope of employment. Courts have ruled that employers are liable for the negligence of their employees even though the employer did nothing to cause the plaintiff's injuries. The employee's negligence is imputed to the employer, whose liability rests on the master-servant relationship. Under scope of employment, the security/FSD should list the responsibilities of each person employed by the organization.

NEGLIGENCE AND LEGAL LIABILITY

A foreseeable risk confronting almost every person and business is being lax in reviewing their insurance policies and failing to correct deficiencies. The basis of the risk is the liability that may be attributed to a party responsible for the harm or destruction to individuals or their property. Courts will rule that for some risks, the foremost predictable loss can be precisely calculated. However, regarding legal liability, the loss depends on the severity of the injury or loss and the sum the jury awards to the injured party.

As the security/FSD, you must look at the facility, your employment practices, training practices, and fire operation plan to ensure that you are doing everything legal in order to avoid a negligence suit. "Liability insurance is rarely concerned with the legal penalties resulting from criminal behavior or intentional torts, for it would be contrary to public policy to protect an individual from the consequences of the intentional injury he or she inflicts."[62] The three main questions asked by courts regarding liability are

1. Was there negligence?
2. Was there actual damage or loss?
3. Was the negligence the proximate cause of the damage?

Many doctrines relate to the law of negligence and are found in statutes. However, the primary practice has been to define negligence through the theory of common law. That theory states, "The basic principal of common law is that most people have an obligation to behave as a reasonable and prudent individual would. Failure to behave in this manner constitutes negligence, and if this negligence leads to an injury of another, or to the damage of another's property, the negligent party may be held liable for the damage."[63] Negligence can also be interpreted as the inadequacy of an individual or corporation to exercise a suitable degree of care required by the situation.

A case with which all FSDs and organizations should be familiar concerns liability in the death of a fireman. This is also known as the *fireman's rule*; it is a long-standing and well-respected principal of tort law that has been part of our legal system and tradition for many years. The case was decided in May, 1988, in Chicago, Illinois, by the Appellate Court of Illinois. It involved Eileen Coglianese, special administrator for the estate of Edmond Coglianese, deceased, and The Mark Twain Limited Partnership, beneficiary under Trust No. 23904.

Edmond Coglianese was employed by the City of Chicago, Illinois, as a firefighter. He died while fighting a fire at the Mark Twain Hotel in Chicago. Eileen Coglianese, his wife, sued the Mark Twain Limited Partnership, manager, and operator of the hotel for the wrongful death of her husband. She alleged that the hotel had caused her husband's death by failing to comply with the building and fire prevention codes for the City of Chicago. She claimed that the interior walls were not fire resistant and were highly combustible. During the fire, it was alleged that the walls burned rapidly, causing black smoke, soot, and noxious gases to be released. Edmond Coglianese, who was wearing his breathing apparatus, suffocated to death from the smoke and noxious gases released during the fire.

The trial court applied the *fireman's rule* and dismissed the case against the hotel operator without a trial. The trial court acknowledged that harm from fire is a reasonable risk of a firefighter's occupation and they further defined the *fireman's rule* as follows:

> While a landowner owes a duty of reasonable care to maintain his property so as to prevent injury occurring to a fireman from a cause independent of the fire, he is not liable for negligence in causing the fire itself.[64]

The courts have determined that the task of a firefighter is to deal with fires. Therefore, firefighters must weigh the consequences commonly identified with this job when they agree to employment. A landowner is liable when a firefighter is exposed to undue hazards beyond those affiliated with fighting fires.

The court also stated that, ordinarily, a violation of the municipal code regulating fire safety precautions would be proof of the hotel's negligence. Why? Because the hotel failed to execute sound judgment. However, this is not so in this case, because the hotel owed no duty to execute sound judgment in deference to a fire fighter, except to hazards unrelated to fire fighting. Yes, a municipal code was violated, but the hotel owed no duty. Therefore, the hotel could not be liable for negligence. The court of appeals upheld the trial court's dismissal of Mrs. Coglianese's action against the hotel for the wrongful death of her husband.

In any litigation involving negligence, the circumstances in any given case are reviewed along with the parties and their actions. In general, the courts rule in favor of negligence to anyone who suffers impairment as a consequence of a person's or corporation's breach of duty, even if the careless party could not have foreseen a risk of harm to someone because of their behavior.

To make a determination of what constitutes negligence, the courts apply to what is referred to as *the prudent man rule*. This rule attempts to determine what would have been the sensible course of action under the circumstances.

Usually when a fire or a catastrophe occurs, and death and injuries follow, many law suits are filed. To avoid this event, it is beneficial for the FSD to meet with the corporation business manager every 4 months and review the various insurance policies. Your input and knowledge may save the company thousands, even millions, of dollars if a negligent law suit can be averted.

In a case that was decided in December, 1982, the Supreme Court of New York ruled that a corporation cannot recover lost profits and other damages caused by a wrongful death of a corporate manager in a hotel fire.

The case *Arrow Electronic, Inc. v. The Stouffer Corporation, et al.* stemmed from a fire at the hotel that killed employees of Arrow Electric Company. On December 4, 1980, shortly after 10:00 A.M., a fire broke out on the third floor conference wing of the Stouffer's Inn of Westchester, in Harrison, New York. The fire spread quickly and took the lives of 26 individuals. Thirteen of the dead were employees of Arrow Electronics, Incorporated. Arrow Electronics sued the owners of the hotel and William L. Crow Construction Company. The construction company had been retained by the hotel to plan, design, and construct the hotel.

In October, 1980, Stouffers agreed to rent guest and meeting rooms to Arrow Electronics for their senior-level annual budget meetings. Arrow Electronics alleged that the hotel implied the rooms were suitable, safe, reasonably free of fire hazards, and reasonably equipped with devices to minimize the dangers associated with fire. In the suit, Arrow Electronics sought compensatory and punitive damages. They argued that the deprivation of its management; costs of recruiting new employees; loss of record books, and papers; death benefits paid to the families of the deceased; and irrevocable revenue were the reasons for the punitive damage suit.

The defendants all moved to dismiss the claims with the exception of the claim for the loss of books, records, and papers, which the defendants agreed was a practical claim.

"The trial judge examined the status of the law on this issued in New York, and ruled that Arrow Electronics was in essence attempting to pursue a wrongful death claim. However, claims for damages for wrongful death belong exclusively to the personal representative of the decedent, according to a New York statute. Furthermore, limiting recovery to the personal heirs of the decedent is the wise course, according to the court, "if recovery could be had by all who are injured in fact, the resources available to compensate . . . the survivors . . . would be diluted."[65]

The trial court dismissed Arrow Electronics' claims for damages or irrevocable revenue, as well as other compensatory and punitive damages resulting from the deaths of the employees.

AVOIDING NEGLIGENCE AND LIABILITY

As the security/FSD, you must continually focus your attention on preventing an injury or a potential hazard from occurring. While employed in the capacity of security/FSD, you are an agent of that corporation. It is your responsibility to establish and maintain a sound security and fire operation plan.

In most instances, you will be employed by a corporation or landlord. Property owners have the right to do what they want with their property. However, the owner or landlord is obligated to the individuals who enter their property. A degree of care must be exercised so that individuals who enter the property will not be willfully harmed. Common law recognizes four classes of individuals with differing degrees of care due to them—trespassers, licensees, invitees, and children.

Trespassers

Trespassers are any persons who enter or remain on the land of another person without that person's consent. Courts have ruled that the owner of the land is under no obligation to guard against injury of a trespasser and is not liable if trespasser's injure themselves unless, of course, an unjustified risk of injury or harm to the trespasser is created by the property owner, such as installing traps to cause injury to the trespasser.

Trespassing minors and discovered trespassers are exempt from this rule. Once a trespasser has been discovered, the landlord must exercise ordinary care regarding the trespasser's safety.

As the security/FSD, it is your responsibility to post signs throughout the facility stating the organization's position on trespassing. The signs should read, *Private Property, Trespassers Will Be Prosecuted To The Fullest Extent Of The Law.*

Licensees

Licensees are persons who enter the property with the knowledge and permission of the landlord. The landlord grants permission to the licensee to perform an act that the licensee could not legally do without permission. The licensee is an individual who is neither a customer, servant, nor trespasser. The licensee does not have any contractual relation with the landlord.

The property owner must avoid intentional harm to licensees and, in addition, must warn licensees of hazardous conditions that may affect them.

If the property for which you work utilizes guard dogs, be sure appropriate signs are posted. The signs should read, *Private Property, Guard Dogs Patrolling the Grounds*. By this action, you have taken the necessary precaution to warn a licensee visiting the property of a potential hazard.

Invitees

Invitees are individuals who enter the property of another by the other's invitation or request. In tort law, occupiers need not ensure the safety of invitees, but they are obligated to provide reasonable care for them from latent defects on the premises that may cause them injury or harm.

A business invitee is an individual who is invited to another person's property for the sole intention of conducting business. The inviter must take rational care to safeguard the invitee from injury and harm. Classifications of invitees are customers, letter carriers, delivery people, workers, trash removal workers, contractors, subcontractors, and visitors.

Property Owner's Obligations

The property owner is obligated to inspect and discover the presence of *natural* and *artificial conditions* or activities that carry any risk of potential injury or harm. If a potential hazard is uncovered, the property owner should caution invitees of these perils and/or make them safe. Any potential hazard that could cause harm to an invitee is a possible source of liability.

Children

The property owner must incorporate a high degree of care in protecting children who enter the premises. As the security/FSD, you must realize that, in past cases, the courts have ruled that the property owner must protect children from themselves, regardless of their status as trespassers, licensees, or invitees.

Regarding children and negligence suits, there is a term with which the security/FSD should become acquainted. It is *the doctrine of attractive nuisance*. This doctrine pertains to tort law. Property owners who maintain a precarious instrument on their premises, which is apt to tempt children, have an obligation to reasonably safeguard those children against the dangers of that attraction. If your facility has a playground or an activity area, and unauthorized children enter your property, you must take precautions to safeguard the trespassing children from being injured.

The doctrine of attractive nuisance applies to a negligent case when the child is so naive as to be unable to recognize the danger involved. The basis of this liability is generally held to be nothing more than the foreseeability of harm to the child.

A building under construction or renovation is an explorer's delight to children. As the security/FSD, it is your responsibility to apprise management of a potential hazard that could invoke the attractive nuisance doctrine. If you take the extra precautions of securing all means of ingress and egress, post No Trespassing signs, and patrol the building, you are taking foreseeable precautions of safeguarding that property.

Property owners have many legal and ethical obligations. As an employee, and in your capacity as the security/FSD, you can be considered an agent or a high managerial agent of your company. Your decisions pertaining to security, fire safety, and loss prevention are vital to the operation. Short- and long-range planning regarding security and fire safety should be reviewed by you or a member of your staff on a daily basis. By having a general knowledge of the elements that constitute negligent liability and by attempting to avoid negligent liability law suits, you are taking foreseeable and practical measures to eliminate hazards in this area. Always consult the organization's attorneys prior to answering any legal queston or subpoena. Never authorize any written reports to leave your office without first consulting the organization's attorney. Never allow an employee to be interviewed by an insurance representative or an attorney without consent from your organization's attorneys. If you or a member of your staff is subpoenaed, first review the case and all documentation with corporation council. It is not a crime to review a case that happened 2 years ago. It is beneficial to your case if you or your staff present themselves as professionals who acted in a prudent manner.

FIRE PREVENTION: AVOIDING LIABILITY

Fire prevention and fire safety in any property can be a nightmare if training, education, and measures of prevention are not incorporated into a fire safety plan for your facility. Like most security/FSDs, a heavy emphasis is placed on security and the protection of life and property. Yet, with proper training and direction, your personnel should be able to fight, contain, and extinguish the fire, thereby preventing loss of life. If your staff is not properly trained and loss of life occurs, you can be assured a lawsuit will be forthcoming.

Multiresidential properties, hotels, hospitals, malls, office buildings, and retail establishments differ in size, design, structure, classification, and occupancy. These differences must be incorporated in your campaign as security/FSD to educate all individuals in fire prevention and safety. You must convince them that fire prevention is everyone's responsibility.

In most multiresidential properties, the board of directors, tenants association, or the property managing agent is responsible for implementing specific fire prevention programs. These committees are usually reactionary when it comes to decision making. They act only after a serious incident has occurred, rather than trying to prevent it in the first place; this is a tremendous liability risk.

The initial step in your fire prevention program is to educate everyone affiliated with the facility in fire prevention. The NFPA suggests the following process for developing a fire safety education program:

I. Initial Planning
A.   Establish responsibility and support
B.   Form a planning team.
C.   Identify local fire problems.
D.   Define goals and objectives.

II. Design and Implementation
A.   Conduct audience/market research.
B.   Develop program strategies.
C.   Develop action plans for program objectives.
D.   Write a program proposal.
E.   Prepare teaching aids and train instructors.
F.   Conduct pilot tests.

III. Evaluation
A.   Provide program documentation.
B.   Determine effectiveness.
C.   Revise action plans and objectives.[66]

Any type of fire at any facility can interrupt the daily activity associated with that facility and create chaos for residents, employees, and visitors. The primary objective of a sound fire prevention program is to protect lives and prevent fires from starting. Careful, systematic, and intelligent fire inspections are good fire prevention measures. These inspections are the backbone of an effective fire prevention program. If conducted regularly and effectively, they can reduce the loss of life, save the facility lost revenue, and reduce liability.

Inspections not only prevent fires, but also afford you and your staff an opportunity to evaluate the property and make contingency plans in the event of a fire. Your three basic objectives are to

1.   stop fires from starting
2.   prevent fires from spreading
3.   comply with the fire/building ordinances in your city.

Many facilities use fire patrols or watch clock tours for fire prevention. A watch clock tour is when the fireguard patrols the property with a watch clock or similar electronic device and checks in at specific stations located throughout the facility, such as in boiler rooms, elevator rooms, and maintenance shops.

Many organizations have the fireguard patrol from 10:00 P.M. to 7:00 A.M., but a fireguard can conduct these tours 24 hours a day. As the FSD, you or a member of your staff must review the fireguard's clock tape or electronic device daily to ensure the tours are conducted properly and in a timely manner.

It would be wise to purchase a second watch clock or electronic device in case of damage to or failure of the original. Many retailers selling these devices will rent a loaner device while your original is being repaired. You would be negligible if it was determined that you failed to conduct fire patrols because of a faulty device. If your facility uses a watch clock with keys, it is best that you order a sufficient supply of keys to replace any keys that are removed or lost. Replace all keys if you discover they have been removed.

Organizations with lone or secluded building sites are constantly searching for means to consolidate control functions to a single location, generally for cost reduction and accountability purposes. Some security/FSDs feel that security and fire safety should be centralized; others disagree. Your organization's administration must make this determination.

Alarm companies will provide transmitter and receiver modules that link remote and central sites. This centralized method of communicating alarms and signals not only cuts costs, but also provides your department with the opportunity to widen its sphere of control when one or more sites are to be monitored. By utilizing a reputable alarm company to monitor security and fire safety you are taking precautionary actions and reducing liability.

If your facility has an indoor garage with self-closing doors, be sure your fireguard inspects the garage hourly. This ensures the garage is patrolled from a security and fire safety standpoint. Most indoor garages have buckets placed every few feet. These buckets should be filled with dry sand, which is an agent for combating and containing the spread of fires, and use them in the event of a fire. However, most people use these buckets as trash receptacles. Good fire prevention means checking these buckets for litter and proper placement.

If your facility has a sprinkler system, you should know when the system was last tested and where the sprinkler shut-off valve is located. In the event of smoke or fire, the sprinkler system should activate, discharge water, and extinguish the fire. Don't place stock any closer than 18 inches to the sprinkler pipe. The FSD should know the origination of the facility's water supply, its adequacy, and the water pressure. You can obtain this information from the building operations department. If your building operations department is unsure, check with the local fire department or the company that installed the sprinkler system. If this information cannot be determined, hydrant flow tests should be made by the local fire or water departments to ascertain if pumpers can access water from nearby hydrants without diminishing the flow to a point at which the sprinkler system would not function. It is the responsibility of the FSD to ensure that the sprinkler system is operational and that certain tests have been conducted to ensure compliance. If it is determined that the sprinkler system was faulty and did not perform as expected during the course of a fire, your organization can be liable for failing to inspect and test the system as required.

Most municipalities mandate that sprinkler systems or standpipe systems be inspected by a qualified person within your organization. As the FSD, you can appoint one of your fire brigade members to this position. Some municipalities require the candidate pass a written examination. If this individual passes the examination and meets the requirements of the local municipality, she may be issued a certificate of fitness from that municipality. This person is then responsible for the record keeping and inspection of the sprinkler and standpipe systems. All buildings possessing sprinkler systems should be inspected on a monthly basis.

There are three types of sprinkler systems with which a security/FSD should be familiar—*automatic wet, automatic dry,* and *nonautomatic* sprinkler systems. An automatic wet system means that all pipes and sprinkler heads contain water or other approved liquids for fighting fires. An automatic dry system means that air is in the sprinkler pipes and heads. An automatic dry system differs from the nonautomatic system in that in the event of a fire it would take time for water to go through the pipes of a nonautomatic system and begin discharging water or other liquids. A nonautomatic system means that there is no water or other liquids in the sprinkler heads or pipes and the water supply source is the local fire department connection. It is imperative that you implement checks and balances in your weekly or monthly fire inspection plan to avoid liability. Check these systems weekly to ensure they are operational; have a qualified employee inspect the pipes and heads for deficiencies.

Check air and water pressures of the dry pipe system and document that the operating pressure is being maintained. Check control valves. Be sure these valves are in the open position and the seals have not been detached. During freezing weather, check the dry pipe valve closets to ensure that they are operational and suitably heated. Check principal water levels of the dry pipe system. Replenish the water or other liquids as required. Survey the fire department's Siamese connections to document that the caps are in place, chain connections are unbroken, threads are undamaged, and the openings have not been impeded. Keep accurate records of all inspections. In the event of a fire at your facility, these inspections will eliminate negligence on your behalf as a high managerial agent.

Avoiding liability as the security/FSD is of paramount concern to you and the organization. Creating and establishing a sound fire prevention program for your facility ensures safety and educates all persons affiliated with the property of potential fire problems. Work with the local fire departments in fire prevention. Many municipalities have a fire prevention department. Find out if your city has one. If it does, meet with a representative for advice. According to the NFPA, "In some departments, inspection personnel from the fire prevention division have the sole responsibility to conduct fire prevention inspections. The objective of their inspections, like those of the general inspections performed by fire companies, is to locate conditions that violate the fire codes that may cause fire or endanger life and property."[67]

Good fire prevention in any facility begins with a fire operation plan. The plan should evaluate the entire facility and include individual floor layouts. You must have an idea of the general population of the facility (e.g., children, elderly, visitors, employees, and the disabled). The plan should incorporate hours of accessibility and to whom the facility is available.

Once an evaluation is completed, map out primary and alternate exit routes individuals can use during a fire. Get the property or building manager's assistance. Examine diagrams or blueprints of the property for each building and floor. Keep these floor plans at your operations center. In the event of a fire, the fire department will need them.

Be familiar with window locks and keys for your premises. The local municipalities in your city have specific guidelines and procedures regarding this area. The mandates usually state that in all high-rise office buildings, which have windows equipped with key-opening locks, a minimum number of keys must be located at the fire command station for use by the local fire department. It is a good idea to determine your municipality's mandates in this area. Document all inquiries you make with the municipality.

SUMMARY

The primary concern of security/FSDs is to determine their objectives, identify and evaluate risks, then implement good security and fire safety procedures. There is a tremendous responsibility associated with the position of security/FSD. If you have security and fire systems in place or are going to incorporate them into a facility, be familiar with the possible liability to which you may expose an organization.

Once the risks have been determined in the areas of security and fire safety, act on them by taking the appropriate measures to correct the deficiencies. Specific risks, because of the severity of the potential loss, demand more attention than others. In most cases, there will be a number of hazards that are equally demanding. Bring potential high risks to the attention of administration. Request that they act to remove the risk as the company can be held liable.

Familiarize yourself with your municipality's local fire and building ordinances. Mandates change frequently and it is imperative that you know how and why these changes affect your organization.

The three areas of liability of which a security/FSD should be aware are—criminal, strict, and vicarious liability. These areas of liability have a direct impact on FSDs, their staffs, and the organizations. Also realize that there are four areas of culpability in which courts determine cases of criminal and tort law—intentional, knowing, reckless, and criminal negligence. If it is determined that an individual or corporation acted in any of the aforementioned manners they can be held liable for their actions or inactions.

It is crucial that you implement weekly or monthly inspections using checklists. These lists will help you assess your organization's security and fire safety systems. If these practices are in place along with good housekeeping procedures, they will aid you and your organization in the event of a criminal negligence lawsuit. All investigative reports should be filed for at least 5 years. All reports, inspections, and services concerning fire safety should be kept in a special area and be readily accessible to you and the fire department.

Be familiar with your organization's insurance coverage and how it impacts your department. The organization's business agent usually holds all policies regarding insurance. Inquire, from time to time, if the policy needs updating or if insurance will decrease if certain security or fire systems are purchased.

In determining negligent liability, the courts have ask three specific questions:

1. Was there negligence?
2. Was there actual damage or loss?
3. Was the negligence the proximate cause of the damage?

If the courts determine that negligence was involved, you, your staff, and the organization can be held liable. It is imperative to remember that FSDs and their staffs are also liable for any tortious conduct.

In *Coglianes v. The Mark Twain Limited Partnership*, the fireman's rule was discussed. It is a case worth reading and a rule worth knowing. The case prompted the question, Can a corporation be held liable for the death of a fireman when the fireman is attempting to fight a fire that a corporation may have been negligible in starting or preventing? The fireman's rule pertains to most municipalities. Inquire from your insurance carrier if they have heard of this rule and how might it apply to your organization.

As the security/FSD, protection of your organization's property from intruders is an area of vital concern. You must be familiar with the four classifications of individuals who enter your property—trespassers, licensees, invitees, and children.

Ensure that an effective security and fire operation plan is in place. This plan is the primary means of ensuring the protection of life and property. Establish checks and balances to avoid negligence and liability, which can be achieved through good fire prevention methods.

After you evaluate the risks associated with the property, take corrective measures to eliminate those risks. Never place yourself or the organization in a position of being sued for negligence. Avoiding liability is everyone's concern.

9

Developing and Implementing a Fire Operation Plan for Your Facility

Each and every structure is designed, built, and constructed differently from another. Municipalities give different types of building classifications to the various buildings throughout the city. Hotels, motels, office buildings, schools, industrial plants, mercantile shops, malls, hospitals, multiresidential homes, single-family homes, churches, and high-rise office buildings are all designed differently from one another.

How you design and implement a fire operation plan depends on the building classification, need, and economics. The NFPA life safety codes set specific criteria pertaining to various occupancies. Standards established by the NFPA Life Safety Code include defining a common path of travel, travel distance to exits, methods of determining egress, provisions for arrows on exit signs, specification of handrails, and calculating seating and aisles in assembly occupancies, to name a few. How these criteria affect your facility should be thoroughly researched.

New York City requires that a fire safety plan must be submitted in writing and approved by the New York City Fire Department. They will make the final determination regarding your facility and fire safety plan. Ask your municipality for their requirements.

The fire safety plan is part of your fire operation plan. The fire operation plan should state the purpose and objective of the plan, which is to establish a method of systematic, safe, and orderly evacuation of an area or building by and of its occupants, in case of fire or other emergency, in the least possible time. The occupants should be ushered to a safe area via the nearest, safest means of egress. The use of available fire alarms and fire suppression systems are vital to achieving your purpose.

The objective of the plan is to provide fire safety education. It is vital that FSDs and their staffs establish programs for all employees, occupants, residents, visitors, and guests to ensure the prompt reporting of a fire or smoke condition. The rapid response to activated fire alarms and the immediate initiation of established fire safety procedures to safeguard life and property, and contain the fire until the arrival of the fire department are also paramount objectives. Fire safety objectives must be clearly defined. "These objectives describe the degree to which the building should protect its occupants, property contents, continuity of operations, and neighbors. The objectives should be quantified wherever possible, rather than stated in broad or general terms."[68]

Chapter 2 discussed the qualifications and requirements of FSDs and their staffs. It is essential for the FSD to choose qualified candidates to implement a sound and effective fire operation plan. The fire operation plan should incorporate the duties and responsibilities of the FSD and her staff in the event of a fire.

THE FIRE OPERATION PLAN

As per Chapter 3, the fire operation plan should be placed in a looseleaf notebook for the purpose of adding or deleting pages as required. The document should be clearly labeled *Fire Operation Plan*. The inside cover of the notebook should contain an index that lists the 11 phases of the fire operation plan:

    I. Fire Safety Plan
   II. Personnel Qualifications
  III. Personnel Selection
  IV. Personnel Duties
   V. Personnel Certification
  VI. Inspections
 VII. Equipment
VIII. Fire Drills
  IX. Coordination With Other Departments
   X. Record Keeping
  XI. Post-Fire Analysis

Each of these phases is critical for an effective fire operation plan. As the FSD, you should become familiar with all eleven phases of the plan and make changes to the plan as needed.

Your ultimate goal is to have the design and contents of your fire safety plan approved by your municipality's bureau of fire prevention. Do not be discouraged if your municipality rejects your initial fire safety plan. Continue to work on it and incorporate their suggestions, so that, ultimately, your plan benefits your organization operationally and economically.

I. Fire Safety Plan

Page one of your fire safety plan should be entitled as such. There should be 11 subsections in your fire safety plan. *Section 1* should have your building address, the name of the facility, and the facility telephone number. *Section 2* should contain a statement that outlines the plan's purpose and objectives. From this point on, it is crucial that you keep your format simple and understandable. *KISS*, keep it simple stupid, should be your guideline. The fire safety plan must be understandable to your employees and the fire department. If it is not, it can cost lives.

Section 3 pertains to the FSD and her duties and responsibilities. List the name of the FSD, her regularly assigned employment duties, and the location to which she is assigned. List the methods of contacting the FSD in the event of a fire (i.e., radio (two-way communication); fire detection, alarm and communication systems, or telephone). Also list the normal working hours of the FSD (e.g., 0800 to 1600 hours Monday through Friday). The FSD is responsible for the following:

1. Maintain and operate a fire safety program that is in accordance and compliance with the municipality's rules and directives pertaining to the fire code.
2. Maintain and operate a sound fire prevention program.
3. Develop and administer a sound fire training program for members of the security and fire safety staff, and the organization who are designated members of the fire brigade.
4. Develop, implement, and ensure that monthly fire and evacuation drills are conducted; evaluate the effectiveness of these drills.
5. Maintain the accuracy of the records of all fire and evacuation drills conducted.
6. During fire conditions, be prepared to advise the fire department officer in command of control and operation of air conditioning and/or mechanical ventilation systems. Inform the fire department officer (city employee) of any other building conditions that may affect the containment or extinguishment of a fire.
7. Evaluate individual floor layouts, the population of the floors during peak hours, and the number and kinds of exits available. Determine the most expeditious escape routes depending on an occupant's location in the building. Calculate alternate routes as well.
8. In the event of a fire condition, fire drill, or fire alarm, ensure that the fire department is aware of the condition by phoning them, pulling the nearest alarm or activating your fire alarm system.
9. Direct evacuation procedures until the arrival of the fire department.
10. Deploy members of the fire brigade to their specific job assignments during a fire condition.
11. Ensure that the fire floor and the floor *above* the fire floor has been evacuated.

12. Provide the fire department's command officer with information pertaining to the severity, location, and nature of the fire condition.
13. Provide the fire department's command officer with a copy of the building's fire safety plan. Keep this plan at the fire command station.
14. Provide the fire department's command officer with the appropriate floor diagrams.
15. Complete the fire command station data sheet, fire report, or fire drill record as required.
16. Ensure that proper entries are made in the security log book.
17. Ensure that appropriate incident reports are prepared and that the information contained within them is accurate.
18. Review the effectiveness of the fire safety plan during and after fire and alarm conditions, as well as during fire drills.
19. Implement revised procedures predicated on the results of the aforementioned evaluation.
20. Ensure compliance with municipality mandates regarding fireguard patrols. Review the watch clock tapes to ensure patrols are being conducted in a timely fashion.
21. Establish and conduct monthly fire inspections of the facility. Review the fire inspection checklist to ensure compliance.
22. Establish and maintain a working relationship with the fire department and fire prevention bureau.
23. Enhance personal knowledge of fire prevention through membership in the NFPA. Study materials provided by the NFPA and local fire department.
24. Become acquainted with the fire, building, and life safety codes. Become familiar with the organizations that enforce and revise the codes.
25. Establish and maintain a rapport with other departments in the organization and keep them apprised of the codes pertaining to fire safety.

Section 4 should contain the duties and responsibilities of deputy FSDs (DFSDs). List the names of the DFSDs and the locations to which they are assigned. Indicate the method of contacting them in the event of a fire (i.e., by radio (two-way communication); fire detection, alarm, and communications systems; or by telephone). List their normal working hours. If your department has a budget for more than one DFSD, it is best to employ additional deputies.

The duties and responsibilities of the DFSD are identical to the FSD, with the exception of items one, two, three, and five listed on the preceding pages. DFSDs are vital to an effective and efficient fire operation plan. If they critique their fire drills and evacuation procedures in accordance with the requirements set forth by the city's fire department, they will make their organization look professional and organized. DFSDs should also distribute policies pertaining to fire safety and fire prevention to other building service employees.

As your department grows, so will the responsibilities associated with fire safety and fire prevention. Selecting qualified DFSDs will only benefit you and the organization. If your facility is occupied and operational 24 hours a day it is best to appoint DFSDs to also patrol the building 24 hours a day. By always having a DFSD on duty, you and the organization are assured that a competent individual is available to handle any fire emergency.

Section 5 pertains to the fire brigade and their responsibilities. Display an on-call list (See Figure 9–1) of the various members of the fire brigade. Define how they are selected and how they are notified when they are not at their regular location. Describe the specific duties of each member.

"Most industrial firms train employees to use first aid fire fighting equipment such as in plant standpipes, hose and fire extinguishers. Thorough training of this type has contributed to a major reduction in fire loss and of life in industrial occupanies. Although initial fire fighting measures primarily protect property, they also protect lives. There is no major threat to life if fire spread is restricted."[69]

Use the following criteria to select qualified fire brigade members:

- their familiarity of the facility
- their mechanical aptitude
- their familiarity with building operating systems
- their availability

2400–0800 Hrs	0800–1600 Hrs	1600–2400 Hrs
	Members Assigned to Assist in Evacuation	
_____	_____	_____
	Members Assigned to Control Small Fires	
_____	_____	_____
	Alarm Box Runners	
_____	_____	
	Backup Alarm Box Runners	
_____	_____	_____
	Communication Officer	
_____	_____	_____
	Members Assigned to Floor Below Fire Floor	
_____	_____	_____

Figure 9–1 Organizational chart for the fire brigade. It is the responsibility of the FSD or DFSD to assign and deploy members of the fire brigade to where he/she feels it necessary.

Again, you must remember that if your department is understaffed, it is necessary to select other employees as members of the organization's fire brigade. Accomplish this by seeking the input of other department managers. Ask them to select qualified employees from their departments to be assigned to the fire brigade.

The type of chart displayed in Figure 9–1 will allow you to organize and coordinate efforts within your department and with other department heads. Listed hours commence at 2400 hours and are subdivided into 8-hour segments. Spaces are provided for the FSD to incorporate the names of qualified fire brigade members.

You may want to list the responsibilities of each subgroup. Members assigned to assist in evacuations will report to the fire floor and will aid and assist any individuals who may be trapped on that floor. It is important to remember that the fire floor may be in flames or may be emanating excessive smoke. Do not attempt evacuation if it endangers your life or the lives of the individuals who have already evacuated to a safe location.

Members assigned to control small fires have a difficult job. A *small fire* is a fire or smoke condition that can be extinguished with a fire extinguisher and will not spread to adjacent areas. Notify the fire department when you observe a fire or a smoke condition. The members who are assigned to control small fires shall go to the fire area with fire extinguishers, two-way radios, and sufficient manpower to combat the fire or smoke condition.

Members assigned as alarm box runners have one specific duty—vacate their assigned post, proceed to the nearest alarm box as quickly as possible, and pull the alarm box lever. The alarm box runner should remain at the alarm box until the fire department arrives or she is advised by the FSD to return to the facility. In most municipalities, the fire alarm box is located on a street corner. The street corner on which an alarm box is located is usually identified by a colored light (approximately 25 feet above the ground) on a light pole, so it can be observed from a distance. In New York City, these lights are orange. The fire alarm box is usually within 10 yards of this pole. The backup alarm box runner is utilized in the event the primary alarm box runner is unavailable.

The communication officer is esssentially the reporter. This officer will arrive at the floor below the fire and inform the FSD, at the command center, of what is happening. The communication officer will advise the FSD of the severity of the fire, and inform the FSD whether or not evacuation procedures have begun and whether or not people remain trapped on the fire floor. The FSD, in turn, will issue this information to the fire department on their arrival.

Fire brigade members who are assigned to the floor below the fire floor should respond there with the primary intent of assisting the evacuation officer. These members should be in two-way communication with the FSD and should remain at their location until the arrival of the fire department. At that time, it will be determined by the FSD or the fire department whether or not these individuals should be relocated.

It is imperative that each member of the fire brigade become familiar with the fire safety plan and their responsibilities. During a fire or smoke condition, the FSD has limited time to relay instructions. If the fire brigade is properly trained and familiar with their responsibilities, the FSD and the fire department will have an easier job.

Section 6 should provide simple, written instructions for employees, residents, occupants, visitors, and guests. (See Figure 9–2) These instructions should outline the need for everyone's participation during a fire emergency, list what individuals should do in the event of a fire or smoke condition, and list what employees, residents, occupants, visitors, and guests should do during a building evacuation.

Issue these instructions to employees via departmental meetings or through the mail in memo format. If your facility is a hospital, hotel, motel, school, office building, or multiresidential property, it is best to try and educate the public as often as possible of your fire safety plan. If your facility has a transient population (i.e., guests stay for a limited amount of time), post these instructions in their rooms so they can become familiar with them. Ensure these instructions are written in several languages. This enables your organization to target a larger population and reduce negligent liability lawsuits.

Section 7 is one of the more critical areas of your fire safety plan. It concerns fire drills and the frequency of these drills. Much thought and planning should be given to this section. List the frequency of the fire drills conducted in your facility. Conduct fire drills once a month on each tour. Many organi-

---

A. Participation

All employees of the facility shall participate and cooperate in carrying out the provisions of the fire safety plan. All residents, occupants, visitors, and guests should become familiar with the facility and follow the instructions of management or the fire department during a fire condition.

B. Alarm Transmissions:

Any individual who discovers a fire or smoke condition should, without delay, transmit an alarm by pulling an interior fire alarm, telephoning the fire department, or pulling the nearest street alarm box. (Indicate the nearest street alarm box location.)

C. Evacuation:

1. Follow the instructions of the fire department or the security/fire safety department.
2. If you are in a room or a closed office that is *not* on fire, do not open the door to any common area (e.g., hallway) until you are instructed to do so by a member of the fire department or security/fire safety department.
3. If you can safely exit your location, do so and proceed to the fire exit or designated fire stairway indicated on the floor diagram posted in your room or by the elevator.
4. Do not use elevators.
5. Keep your children or family members with you.
6. Do not panic or run.

Figure 9–2 Instructions during a fire.

zations use outside agencies to conduct their fire drills. These agencies charge a fee, but also keep accurate records, list any deficiencies found during the fire drill, and give a written synopsis to the FSD.

If your organization intends to conduct its own fire drills, specify how the drills are to be conducted. Fire drills can be initiated by manual pull stations situated throughout your facility and activated by a member of your fire brigade. Have your fire brigade and DFSDs respond to a simulated fire condition and have them carry out their assigned duties. As FSD, you should report to the fire command station and perform your assigned duties.

Accurate record keeping of fire drills is essential. (See Figure 9–3) Records should be kept on file for at least 5 years. The fire drill record allows the FSD or DFSD to review and critique the participation of the department during a simulated fire or smoke condition. The fire drill record as well as a written report should be submitted to your administration after the drill has been conducted. The report should outline what was done, who participated, and what recommendations you can make in regard to the fire drill.

Many fire departments throughout the United States require that a fire drill be conducted approximately every 3 months. It is beneficial to you and

---

Date of Report: _____

Date of Fire Drill: _____ Time of Fire Drill: _____

Personnel Who Participated in Drill:
1. _____ Department: _____
2. _____ Department: _____
3. _____ Department: _____
4. _____ Department: _____
5. _____ Department: _____

Was an alarm sounded for the drill? Yes _____ No _____

Which areas of the facility were affected? _____

_____

Was an evacuation of these areas completed? Yes _____ No _____

If an evacuation was not completed, explain why. _____

Rate the Effectiveness of the Drill:	(1) Good	(2) Fair	(3) Poor
1. Personnel response	1	2	3
2. Occupant, visitor, guest response	1	2	3
3. Personnel familiarization with duties	1	2	3
4. Effectiveness of procedures	1	2	3
5. Speed of evacuation	1	2	3
6. Communication during drill	1	2	3
7. Personnel's familiarity of alarm systems	1	2	3

Fire Safety Director: _____ Date: _____

Certificate of Fitness Number: _____

Figure 9–3   Fire drill record.

your organization to conduct these drills monthly. With a 3-month schedule, personnel become lax, staffs change, and people become less familiar with what they are required to do. Mandating that your facility run fire drills on a monthly basis should heighten personnel awareness pertaining to fire safety and fire prevention.

Stephen Christy, the Director of Safety and Security for Homes For The Homeless, and I put together a comprehensive fire safety plan for his homeless facilities in New York City. After numerous meetings with the New York City Fire Department and the Fire Prevention Bureau, we were able to get our fire safety plan approved. As a result of his memos to various department managers requesting participation from them, Homes For The Homeless became one of the better, fire-safe homeless facilities in the New York City area. Stephen Christy feels that the key to his success was the training of his staff through the utilization of effective fire drills.

Section 8 should detail your facility's fire command station. It should specify the location of the fire command station and to whom access is granted. It must also be adequately lit and accessible by radio and telephone communication. Indicate the areas to which the communication system leads, including elevators, individual floors, boiler rooms, etc. Most facilities have fireman telephones located at the fire command station. If your municipality requires these phones be present, ensure they are installed. These telephones are another component of your fire detection and communication system.

Your fire command station should also contain a copy of your fire operation plan, a copy of your building information data sheet (see Figure 9–4), representative floor plans, and keys to rooms and elevators. This stipulation should be clearly defined in this section of your fire safety plan.

In many high-rise office buildings, hospitals, and hotels, the FSD may assign fire wardens to the floors on which individuals work. This is a voluntary assignment and is an extremely important aspect of fire safety and fire prevention. The fire warden, upon hearing an alarm, should proceed to the elevator and direct personnel, visitors, and guests to use a safe stairwell. If the floor has a phone, the fire warden should contact the FSD at the fire command station. Fire wardens should be part of your fire operation plan. Many FSDs do not utilize the services of other employees effectively. If you choose to use fire wardens, keep them informed of all fire-related information, incorporate them in your fire operation plan, and use them when conducting your fire drills. Fire wardens should become familiar with the facility and, specifically, the floor on which they work. Deputy fire wardens should also be appointed in the event the chief fire warden is unavailable.

This section should provide floor diagrams for each floor of the facility. They should detail the layout of the stairs (indicating those that terminate at rooftops), rooms, corridors, elevators, and standpipe risers. Stairways and elevator banks should be given alphabetical letter designations; individual floors should be given numerical designations. The reason for assigning num-

1. Address of Facility: _____

2. Owner of Facility: _____

   Address of Owner: _____

3. Fire Safety Director's Name: _____

   Deputy Fire Safety Director's Name: _____

4. Does a Certificate of Occupancy exist for the building? _____

5. Building height, area, class of construction, and number of floors: _____

6. Number, type, location, and alphabetical and numerical listing of the facility's fire stairs: _____

7. Number, type, and location of horizontal exits:_____

8. Number, type, and location of elevators: _____

9. List the number of interior fire alarms and smoke detectors as well as their locations. If your facility utilizes heat detectors or manual pull stations, list them here as well.

   _____

10. List the type of communications system your facility utilizes. Include the number and type of portable radios, the number and location of fireman's stations, and the type and location of your emergency public address system.

    _____

11. List type of standpipe system your facility utilizes. _____

12. List the type of sprinkler system, such as dry or wet, and the locations of water supply, the control valve, and the water flow alarms.

    _____

13. List the type, number, and location of special extinguishing systems, such as kitchen *ansul system.*

    _____

14. Indicate the average number of employees your facility employs during day and evening hours. This number should be an estimate. If your facility is open 7 days per week, take this into account as well.

    _____

15. Indicate the total number of handicapped persons that work in or visit your facility.

    _____

16. Indicate the average number of visitors and guests at your facility on any given day.

    _____

17. List the types and locations of the building's service equipment, such as source of building power, type of lighting, emergency lighting, and type of heating systems.

    _____

18. Indicate the type of ventilation your facility utilizes. _____

19. Indicate the types and locations of the facility's air conditioning systems. If it is an incremental system, indicate its location. If the systems are electrical cooling units, list on which current they operate and give their locations. If the systems are air handling units, list on which current they operate and give their locations.

20. Indicate how refuse is stored and handled. _____

    _____

21. List the facility's fire fighting equipment. _____

22. List the types and locations of flammable solids, liquids, and gases. Indicate the location of the master key for these areas.

_____

23. List the types and locations of hazardous material. Indicate the location of the master key for these areas.

_____

24. Indicate whether or not your facility utilizes electrical transformers that contain polychlorinated biphenyls (PCBs).

_____

Figure 9–4   Building information data sheet.

bers and letters is to provide standard identification points in identifying specific stairwells and floors. (See Figure 9–5.)

Section 9 should indicate the various signs your facility uses. Landing signs are an example of the importance of these signs. Landing signs are necessary for all building classifications and should be posted and maintained at all elevators on every floor. This sign should read, In Case of Fire, Use Stairs Unless Otherwise Instructed. Inquire from your city's building department if specific criteria apply to your signs. For example, in most states, the lettering on these signs should be at least 1/2-inch block letters in red and the background should be white. The lettering should be properly spaced and legible. The sign should also contain a diagram showing You Are Here and the location and letter identification of the stairs on that floor. The sign should indicate the escape route for that floor. It is customary for the sign to be at least 10″ × 12″. The sign should be located directly above the call button and securely attached to the wall. The top of the sign should not be more than 6 feet from the floor. Some cities allow you to omit the floor diagram provided that signs containing such diagrams are posted in other areas throughout the floor. In such a case, the size of the landing sign can be less than 10″ × 12″.

Floor numbering signs should be positioned and maintained within each stair enclosure on every floor. The number of the floor should be indicated. Use contrasting colors; the numerals should be one color and the background should be a different color. The sign should be affixed to the stair side of the door.

In hotels and motels, as well as other specified building classifications, signs should be affixed to the back of the door of all rooms that open onto a public corridor. The signs should show the occupant where they are presently located and show the location and letter identification of the exit stairs on that floor.

All signs should be made of metal or other durable material that can withstand a fire condition. "Wood may be used for external decoration only if placed not less than 2 in. (50 mm) from the nearest lampholder or current carrying part. Enclosures for outside use should be weatherproof. All steel

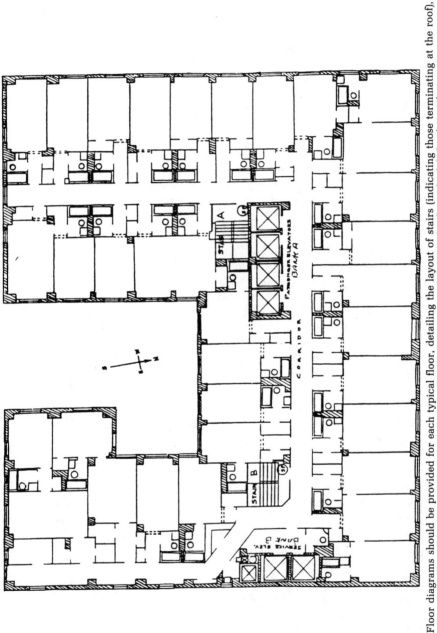

Figure 9-5 Sample floor diagram

Floor diagrams should be provided for each typical floor, detailing the layout of stairs (indicating those terminating at the roof), rooms, corridors, elevators and standpipe risers. Identify stairways and elevator banks with alphabet letter designation. (Reprinted from Hotel-Motel Fire Safety Directors Course Instruction Manual, Part 2, 1986–1987, Fire Science Institute, John Jay College, 444 West 56 STreet, New York, New York. Director Richard Abbott. p. 39.)

parts of enclosures must be galvanized or otherwise protected from corrosion. Signs, troughs, tube terminal boxes, and other metal frames must be grounded in the manner specified in the NEC unless they are insulated from ground and from other conducting surfaces and are inaccessible to unauthorized persons."[70]

Section 10 should outline your fire prevention and fire protection program. List how you will conduct your facility's hourly fire patrols. This can be done using a watch clock or electronic bar code device. Specify when these tours are to be conducted and who will conduct them.

List when your monthly fire drills are conducted and who conducts them. List all fire training programs that you or DFSD will be conducting. If you intend to interface with the local fire department and conduct joint fire drills, outline these procedures in this section.

List the societies you and your staff have joined, and the educational classes in which you want to enroll to enhance your fire safety department. Indicate whether you are a member of the NFPA or any other organization related to fire safety and prevention.

Section 11 should indicate who prepared the fire safety plan, the date it was prepared, and the dates of any revisions to the plan. Any revisions should be included in the fire safety plan and copies should be forwarded to the fire department, indicating those changes that have been made.

II. Personnel Qualifications and
III. Personnel Selection

Chapter 2 discusses the personnel qualifications and the selection of all members of the security department who become candidates for fire safety. As you begin your selection process, it is imperative that you consider the responsibilities of each member of the fire safety department and whether the members of your department meet those qualifications. In most municipalities, tests are given to ensure a candidate's ability. If the candidate you select does not meet the qualifications specified by your municipality and fails the written test, you must select another candidate.

IV. Personnel Duties

The personnel duties you establish for your facility depend on the classification of your building, the type of population at your facility, and the number of employees you have to protect. Hotels and motels will be vastly different from hospitals, and hospitals will be vastly different from multiresidential properties due to populations and classifications.

The NFPA 101 Life Safety Code is an excellent resource for any FSD. In order to establish effective personnel duties for a facility, FSDs should first consult this book. All types of occupancies are discussed as well as general requirements, special provisions, and means of egress requirements. One of the fundamental requirements listed in this book best explains what the 101 Life Safety Code represents. "Every building or structure shall be so constructed, arranged, equipped, maintained and operated as to avoid undue danger to the lives and safety of its occupants from fire, smoke, fumes, or resulting panic during the period of time reasonably necessary for escape from the building or structure in case of fire or other emergency."[71]

As the security/FSD, you must be aware of how building, fire, and life safety codes affect your facility. How would you prepare an effective fire safety plan for a windowless or underground structure? There are special provisions listed in the Life Safety Code regarding these types of structures. "Windowless or underground structures with an occupant load of more than 50 persons shall be protected throughout by an approved automatic sprinkler system in accordance with Section 7–7."[72] The exception to this is existing structures that have an occupant load no greater than 100. Do not be intimidated by different types of facilities. Seek answers to questions regarding fire safety and fire protection. Inquire how your facility may be different or similar to other structures. The duties you outline and prepare for your staff must coincide with the type of facility you are protecting.

V. Personnel Certification

The certifications received by you and your personnel should be listed and placed in your fire operation plan. If your municipality offers a fire safety course, you and members of your organization should take it. Inquire, from your municipality, the requirements for enrolling in this course. Ask if a written examination is given, if a passing grade must be attained, and if the municipality issues a certificate of completion. If your municipality provides these features in its course, send as many employees to the class as your budget dictates.

The classification of your occupancy has much to do with the type of course in which you would enroll your employees. For example, if your facility is a high-rise office building, be sure that your employees attend courses pertaining to high-rise facilities. If your facility is a hotel or motel, be sure the course is designed to educate you regarding fire prevention and fire safety for hotels and motels.

Subsequent to the employee receiving a certificate of completion, the municipality may require an on-site examination. If this examination is required, it is beneficial for the employee to learn as much about the facility as possible. After employees successfully complete their on-site examination, a certificate of fitness may be issued.

VI. Inspections

Fire inspections not only prevent fires but present opportunities to better evaluate and inspect your facility to ensure compliance with fire, building, and life safety codes. Fire inspections should be carefully and systematically planned, and should emphasize fire prevention. Your plan should incorporate how you will prevent fires from starting and from spreading, and how you will ensure compliance with the fire protection, building, and life safety codes.

Familiarize yourself with the local laws and administrative codes of your city. Many times, the inspection of your facility is predicated on its type of occupancy and building classification. Realize that local laws effect smoke-detecting devices, power sources, general requirements for smoke-detecting devices, and the inspection of these devices.

Outline how your inspections will be conducted, who will conduct them and who will review what was inspected. Have members of your department conduct inspections on a daily, weekly, or monthly basis. Use the sample forms presented in the previous chapters to conduct proper inspections. Never underestimate the value and effectiveness of your inspections. Your municipality will be impressed when you specify the purpose of inspections and how you intend to conduct them.

If your facility uses a watch clock or electronic device for recording daily fireguard patrols, discuss this in this section. Keep a separate log book for fireguard patrols. In this log book, indicate the locations of all the stations, and the name of, date, and time the employee made the watch clock tour. Upon employees' completion of their tours, they should sign the log book, indicating the time they began the tours and the time they completed the tours. A supervisor's signature should accompany each entry. This log book should be filed for 5 years. During an inspection by the fire department, it would behoove you to show them these records. Employees should look for the following items during inspections:

- adequate lighting in stairways and hallways
- operational emergency lighting
- accessibility of doors serving as means of egress
- availability and proper spacing of fire extinguishers
- usability of fire extinguishers
- posting of no smoking signs
- operational and properly placed exit signs
- rubbish accumulation or unsanitary conditions
- obstructed sprinkler heads
- improper storage of hazardous materials

Incorporate this check list in your fire operation plan.

VII. Equipment

A complete list of the equipment your facility utilizes pertaining to fire protection and fire prevention should be presented in this section. Discuss your alarm, detection, and communications systems. List their locations and the functions they perform during a fire or smoke condition. If your facility utilizes a fire command station, list its components and describe its function. If your facility has a sprinkler system, list the location, type, and location of the shut-off valve.

Do not forget to list your emergency lighting system, its location and how it is operated. Also detail what type of communication system your facility uses. Under communication systems, list the portable, two-way radios used. Invest in portable electronic megaphones (bullhorns). During a fire or smoke condition, these megaphones are very effective in communicating instructions to persons trapped on fire floors.

Again, your building classification may impact the regulations pertaining to portable fire appliances, including the type of fire extinguisher required (e.g., 1½- or 2½-gallon water-type extinguishers for every x amount of square feet or fraction thereof of floor space). Many municipalities have exceptions to these specifications if the building is equipped with a sprinkler system. List how many fire extinguishers are on each floor and where they are located. Indicate the type of fire extinguisher (i.e., type A, B, C, or D).

Municipalities usually indicate the type of fire extinguishers required as well as the distance of travel between these extinguishers. Ascertain from your municipality's fire department the type of fire extinguisher and distance of travel that apply to your building classification.

Indicate if you have a heavy-duty bolt cutter as part of your equipment; this is an inexpensive investment. In some residential properties, occupants may fall asleep while cooking something or while smoking (thus starting a fire), and their doors may be secured with a regular lock and a chain lock. Building superintendents usually have keys to these apartments. However, many times the chain lock is in place and access is denied. By bringing bolt cutters with you in a fire or smoke condition, you are saving valuable time and energy, and providing easy access to a fire or smoke condition.

If your facility utilizes versatile electric vehicles as part of your security or fire safety department, be sure these are listed. Many times, these electric vehicles are incorporated into a fire safety plan. If the area of patrol is thousands of square feet, it is usually beneficial to have one of these vehicles. List the specifications of the vehicle, including the model number, the vehicle's dimensions, the weight of the vehicle, the wheel base, wheel tread, and type of power the vehicle utilizes. If you use this vehicle to transport fire equipment during fire drills, discuss this in your fire operation plan.

VIII. Fire Drills

In this section, discuss when your facility conducts fire drills and how and where the fire drills are to be conducted. List the municipal authority that requests that fire drills be conducted and accurate records maintained. List the chapter, title, and section of the sanctioning authority you are required to follow.

Define *fire drill* and *evacuation* (e.g., a methodical, safe, and efficient evauation of an edifice, by and of its inhabitants, in the least possible time). Define your objectives for an effective fire drill (e.g., to provide appropriate education and training to ensure the immediate, spontaneous response of security, fire safety, and building operations personnel to the reports or alarms of fire or smoke).

Outline your fire drill procedures. These procedures should be identical to those as followed in the event of an actual fire or smoke condition, except that all duties pertaining to evacuation, fire extinguishment, and containment will be simulated.

"Fire drills in health care institutions are usually conducted as a part of the orientation program for new employees. Later the drills are supplemented with in-service training for the staff personnel, including the emergency procedures. Fire drills in many facilities are conducted once a month on each shift. The training for the drills typically involves instruction and practice for the staff personnel in the various means of moving nonambulatory patients, the procedures for alerting the facility staff, and the method of notifying the fire department."[73]

Health care facilities have been purposely singled out due to the difficult task the security/FSD faces on a daily basis. The procedures FSDs must establish and implement in a health care facility are more difficult than those of many of their contemporaries. Evacuation procedures are of utmost importance in a health care facility. Trying to move or relocate a nonambulatory patient can be life threatening in itself. The FSD should keep four sequential stages of an orderly evacuation of a health care facility in mind—"(1) the manpower supply phase, (2) the patient preparation phase, (3) the patient removal phase, and (4) the rest and recovery phase {Archea 1979}. This approach focuses on the occupants in the fire-threatened area and the patients in or adjacent to the fire area. Removing immediately threatened patients, closing doors to the room of fire origin and to adjacent patient rooms, would be compatible with this four phase approach to evacuation."[74]

The report writing and record keeping associated with the organization's fire drills is mandatory. Appropriate reports must be submitted when a fire drill is conducted. These records must be logged and maintained. A journal must be maintained strictly for recording fire drills. The cover of this bound journal shall be labeled *Fire Drill Record*.

Entries in this journal should be made by the FSD or the DFSD upon completion of the fire drill. The FSD or DFSD must critique the effectiveness of the fire drill and make recommendations when warranted. These recommendations or the ineffectiveness of the drill should be brought to administration's attention via memo. The fire drill record shall be maintained by the organization for a 3-year period and shall be readily available for inspection by the local fire department. When a representative from the fire department reviews the fire drill record, have her sign the journal, indicating her approval.

During a fire or evacuation drill, be sure your elevators provide for *Firemen Service*, which is usually a city law. Firemen Service allows for immediate availability of elevators for fire department use during a fire or smoke condition. It therefore prohibits individuals from using the elevators during a fire condition.

If your facility presently has firemen service controls, additional keys to activate the elevator controls should be kept at the fire command station for the fire department's use. When the fire department arrives, issue these keys to the individual in charge. When you conduct your drills, incorporate the discontinuance of elevator service for a specified bank of elevators.

If your facility does not have firemen service controls, additional preparation and communication is essential. In the event of a fire, all elevators effected by the fire floor should be manually returned to the building's lobby. Unless, of course, the fire is in the lobby. The elevators should not be put back into service until authorized by the fire department.

There are two types of elevators with which you should be familiar—electric and hydraulic elevators. Elevators are classified by their driving means and their use. "A passenger elevator is used primarily to carry persons other than the operator. A freight elevator is used for freight; only the operator and the persons necessary for unloading and loading the freight are permitted to ride. Service elevator is another term used to describe elevators. A service elevator is officially classified as a passenger elevator that has also been designed for the carrying of freight."[75]

If your facility utilizes escalators, incorporate their shut-down and use in your fire drill plan. "An escalator can be defined as a power-driven, inclined, continuous stairway used for raising or lowering passengers (ASME 1984b)."[76]

There are specific requirements established by the NFPA that pertain to escalators and fire safety. "NFPA 101 requires escalators to have their floor openings enclosed or protected as required for other vertical openings. An important exception allows escalators not to be closed in completely sprinklered buildings if the escalators are protected by alternate means."[77]

Elevators, escalators, dumbwaiters, and moving walks must be installed in accordance with the requirements of the *Safety Code for Elevators and Escalators*. If your facility uses any of these mechanical devices for transportation, they must be incorporated in your fire safety plan and your fire drills.

A critique and evaluation of the fire drill must be written by the FSD or DFSD. It should be based on information developed from observation, communication, and participation. The effectiveness or ineffectiveness of the drill must be discussed with the staff as well as reported to administration.

IX. Coordination With Other Departments

This section should be a written statement regarding how other departments are involved with the fire operation plan, their functions and responsibilities, and how they are notified in the event of a fire or smoke condition.

Depending upon your building classification, alphabetically list those departments that are incorporated in your fire operation plan. If your facility is a large department store, list what you have discussed with the other departments pertaining to fire safety.

For example, a sample written statement is, "various departments were present for the fire safety meeting held on _____ at _____ hours. The responsibilities of each department manager and their staffs were discussed. Also discussed were topics pertaining to new employees, designated fire lanes, and illegal parking in fire lanes. Recommendations and alternatives were outlined.

Departments Present

Accounting, Mr. Strong

Building Operations, Mr. Sweeney

Customer Service, Mrs. Mercaldo

Garage Manager, Mr. Phillips

Health Care Representative, Mrs. Sweeney

Loss Prevention Manager, Mr. Oakes

Operations Manager, Mr. Crosby

Personnel Manager, Ms. Harmon

Risk Manager, Mr. Ciribisi

Sales Manager, Ms. Tartagleone"

You can expand or decrease this list depending on the number of employees who attend the meeting. Be sure you circulate a memo stating the date and time of the meeting and the topic to be discussed. Distribute an attendance sheet and have all individuals sign it prior to the beginning of the meeting. Keep these attendance sheets with your other records.

As the FSD, you want as much input as possible from other departments, but you will not get it unless you make them an integral part of your fire operation plan. Your goal is to convince them of the importance of fire safety and their involvement in it. You can accomplish this goal by cutting out articles pertaining to fire and fire safety and making comparisons of those articles to your facility. Explain the importance and the functions of a fire brigade. Never demand their cooperation, ask for it.

Evaluate each department and determine what you want them to contribute to your fire operation plan. Measure each department in terms of staff, productivity, and their knowledge of the facility. After you assess how they meet these criteria, assign a specific area of responsibility to each department. A general exception pertains to the building engineer. He should not be assigned as the FSD because he may have to vacate the fire command station to fulfill his obligations as building engineer. During an emergency, the FSD must always remain at the fire command station. "The building engineer should never be designated as the Fire Safety Director. In the event of a serious fire, the engineer's required position is not compatible with the required position of the Fire Safety Director."[78]

Whenever memos pertaining to fire safety or fire prevention are issued to department managers a copy must be placed in this section of the fire operation plan. If these memos are interim directives, they should be labeled as such. Retain all memos pertaining to fire safety and fire prevention for 3 years.

X.   Record Keeping

Depending on the size and availability of space allotted to your department, records should be maintained for 5 years. Some local laws mandate 3 years, however it is best to hold onto these records, if you can, for a 5-year period. Check with your municipality and inquire how long specific records should be kept.

The safekeeping of records ensures that all data regarding security and fire safety have been addressed and indicates that your facility was in compliance with specific mandates. It also ensures that the FSD and DFSDs were aware of and complied with changes in the fire, building, and life safety codes.

Two sets of identical records should be maintained by your department. As ironic as it may sound, fire may destroy these records, or they can be lost or stolen. By storing a second set of records off-site, you are ensuring that, in the event of a catastrophe, your records will be available for review in a minimal amount of time.

You should advise administration to use protective, fire resistant safes or containers to protect vital corporate records in the event of a fire. Many times, this is an oversight of the FSD. Remember, the organization must continue to operate after a fire condition. Using protective safes or containers for records is one method of safekeeping. "Heavy insulated, massive vaults, safes, and filing cabinets are the traditional way to safeguard valuable records against the effects of fire. The practice of encapsulating the records so that they are fully protected against the greatest potential fire exposure is still viable. Containers are available that will keep the internal temperature and humidity as low as necessary to retain the data on magnetic tape and other temperature sensitive media."[79] These containers should also be airtight to prevent water seepage. Many times, water, not fire destroys vital records. During a fire condition, many gallons of water are used to contain and extinguish a fire. This water can get into record storage areas and destroy records unless appropriate precautions are taken.

"Protective containers for records are rated by tests under standard fire conditions. They are rated according to the time elapsed before the interior of the safe or container reaches 350° F (177° C). This time factor provides a measure of safety, since the ignition temperature of most paper is somewhat higher."[80] (See Table 9–1)

XI. POST-FIRE ANALYSIS

After a fire, it is critical for the FSD to review the causes and effects of the fire. To aid in the FSD's analysis, specific questions and procedures should be in place. After a fire condition, having these procedures available, the FSD can begin to get the facility operational.

The post-fire analysis serves a twofold purpose: (1) the analysis should list specific questions regarding operational capability for the FSD and the administration and (2) the analysis should answer questions ranging from how the fire started to why it started.

In this final section of your fire operation plan, include blank copies of all forms and data you use in your fire operation plan. These blank copies are for your review as well as for the review of your municipality's fire department. Place copies of floor plans or diagrams of your facility in this section as well. If the diagrams are too big, have them reduced by a printing company.

State the purpose for keeping records. The reason for their maintenance is three-fold—they are required by law, they provide administration with necessary information to implement and redesign flaws in the fire operation plan, and they provide administration with information regarding fire safety procedures, the effectiveness of these procedures, the effectiveness of the personnel responsible for these procedures, the evaluation of training programs to ensure an effective fire operation plan.

Record keeping is vital for your fire operation plan. Maintain the following specific records:

- Fire Safety Plan
- Fireguard Inspection List
- Monthly Fire Protection Inspection Checklist
- Fire Operation Flow Chart

Table 9–1   Fire resistance of record containers

Insulated record vault door	2,4, and 6 hrs
Insulated file room doors	½ hr and 1 hr
Steel-plated-vault doors (with inner doors)	about 15 mins
Steel-plated doors without inner doors	less than 10 mins
Modern safes	1, 2, and 4 hrs
"Old line," "iron," or "cast iron" safes, 2 to 6-inch wall thickness	Uncertain
Insulated record containers (files, cabinets, etc.)	½, 1, and 2 hrs
Containers with air space or with cellular or solid insulation less than 1 inch thick	10 to 20 mins
Uninsulated steel files, cabinets, wooden files, wooden or steel desks	about 5 min

Arthur E. Cote and Jim L. Linville (Eds.) (1986) *The Fire Protection Handbook*, Sixteenth ed. Quincy, Mass.: National Fire Protection Association. p. 11–74.

- Building Information Data Sheet
- Instructions During a Fire
- Fire Drill Record
- Monthly Fire Alarm Record Profile
- Incidents Reports

You should design a fire alarm record data sheet for your facility. Any time a manual pull station, smoke detector, heat detector, or sprinkler system is activated, make an entry on the data sheet indicating the date, time, type of alarm, location of the alarm, and any comments you have regarding the activated device. (See Figure 9–6). In this way, you can keep track of the alarms activated in your facility. It will give you as well as administration an indication of the manpower hours alloted to responding to activated alarms, and will also establish patterns of individuals who pull alarms for fun. With accurate record keeping, you can track the days of the week, time of day, and locations at which alarms are activated most often.

This data sheet provides another check and balance in your fire detection system. If a certain smoke detector or pull station continually malfunctions, you can track it and have the alarm company make the necessary repairs to the system. Review and analyze the sheet on a monthly basis. After a few months, you will begin to see patterns develop that will assist you and your department.

Every department utilizes incident reports for their facility. Many facilities design their incident reports to coincide with the facility's type of classification. All incident reports should be labeled *privileged* and *confidential,* and should be individually numbered. The date, time, and page number

	Date	Time	Type*	Location	Comments
1.					
2.					
3.					
4.					
5.					
6.					
7.					
8.					
9.					
10.					
11.					
12.					
13.					
14.					
15.					

*Type of activation can be MS = Manual Station, SD = Smoke Detector, HD = Heat Detector, SS = Sprinkler System

Prior Month's Number of Activations: _____ + or − _____

Day of Week Most Affected: _____

Time/Hour Most Affected: _____

Location Most Affected: _____

Fire Safety Director: _____ Date Reviewed: _____

Figure 9–6   Monthly fire alarm record profile.

corresponding to the entry in your facility's log book should be listed on the top of the incident report.

The incident report is one of the most important pieces of documentation used by any department. It tells an actual story of the events that transpired during a specific time. Incorporate any incident pertaining to fire safety in these reports.

Some corporations use preprinted forms containing incident report data. If these forms suit your facility, use them. If they don't, tailor one to meet your needs. The incident report should contain vital information that

Report Number _____ Date of Report _____ Time of Report _____

Incident Category _____ Page Number _____

Date of Incident _____ Time of Incident _____

Subject's Name _____   Male _____   Female _____

Subject's Address _____   City _____

Subject's Phone Number _____   Subject's Age _____

Suspect or Accused Name _____   Male _____   Female _____

Suspect's Address _____   City _____

Suspect's Phone Number _____   Suspect's Age _____

Incident Reported By _____

Incident Investigated By _____

Department Affected By Incident _____ Manager's Name _____

Was any city agency involved in this incident? if yes, specify.

_____

Write a brief synopsis of the events pertaining to this incident.

_____

_____

_____

_____

_____

INVESTIGATING OFFICER'S NAME _____ ID NUMBER _____

Figure 9–7   Incident report/privileged and confidential.

indicates who, what, when, where, why, and how. (See Figure 9–7) These reports should be distributed on a need-to-know basis. It is a good policy to never allow other departments to have access to security/fire incident reports without first checking with administration.

It is essential that written reports are constructed properly and that facts about the incident are accurately reported. Most security officers are judged, to a great extent throughout their career, by their written and verbal skills. Poorly written reports or unintelligible communications will result in confusion and misunderstanding and will reflect unfavorably on the abilities of the reporting officer.

Poorly written investigative reports stem from three reasons—the writer fails to get as much factual information as possible, the writer draws conclusions in the synopsis rather than getting all the facts, and the writer cannot write concise, well-structured sentences.

SUMMARY

Developing and implementing a fire operation plan for your facility is not an easy task. The purpose of the fire operation plan is to establish a method of systematic, safe, and orderly evacuation of an area or building, by and of its inhabitants, in the event of a fire or other emergency. Your primary objective is to educate people about fire safety and prevention and include qualified people in the fire brigade.

The fire operation plan should be composed of 11 sections—fire safety plan, personnel qualifications, personnel selection, personnel duties, personnel certification, inspections, equipment, fire drills, coordination with other departments, and record keeping. Your ultimate goal is to prevent loss of life and property damage.

10

Hazardous Materials and Fire Safety

Not only are FSDs faced with securing and safeguarding their facilities, but they must also understand that, at any given time, a crisis may develop in which their assistance or guidance may be needed. Let's assume you are responsible for a health care facility. You are responsible for security and fire safety. Another department is solely responsible for safety, but the director of that department is on vacation and his assistant phoned in ill. You have just been notified that a truck making a delivery to the loading dock with unknown material on board just rammed the loading dock and liquid is oozing from the truck. What do you do?

A good FSD will have prepared in advance for emergencies such as this. It is important for the FSD to interface with the director of safety and inquire what steps to take in the event of a hazardous material spill. A special section of your fire operation plan should contain a contingency plan for dealing with hazardous materials.

Fire science has progressed tremendously in studying fire protected buildings, materials used in construction, and occupancy load of these buildings. We have gained tremendous knowledge from fire testing materials, to see how long they burn, the amount and kinds of toxic gases emitted from the material, and which agents to use to fight or contain specific fires.

CHEMICAL SAFETY AND ENVIRONMENTAL COMPLIANCE

As the FSD, you must be able to prepare for and act on any emergency situation. In many facilities, today, chemical safety and environmental compliance must also be taken into consideration. In most facilities, the company will employ a director of safety to be responsible for the storage, handling, and transportation of these chemicals.

Interface with this individual to ensure safety and code compliance. Environmental noncompliance has more costs associated with it as well as

149

more prison time. Your security/fire safety department should be trained and prepared to respond to a chemical spill at your facility. It is essential to contain the chemical spill before toxic fumes are released or the chemicals react to cause an explosion.

View the various chemical agents stored or used at your facility as extremely hazardous. If these chemicals are spilled, they will adversely affect your facility and its continuous operation. If your facility does not have a director of safety, it is your responsibility to coordinate plans to cope with safely storing, transporting, and maintaining control of these materials.

In many instances, management that deals regularly with hazardous materials think it is prepared to handle a spilled hazardous material until a crisis develops. Usually, the first department called is the security department, regardless of whether or not it is prepared to handle the emergency.

Hazardous materials, either stored or in transit, affects surrounding areas as well as their immediate location. As the FSD, insist on carrying out a risk analysis audit (RAA) to effectively measure the scope of the management problem and make recommendations. According to *The Factory Mutual Systems Approval Guide,* the purpose of an RAA is to elicit corporate self-evaluation. An RAA helps corporate management develop sound attitudes about risk exposure, recognize exposures to loss, and, if necessary, seek further clarification about risk from specialists in the loss prevention field.[81] The RAA is, then, an avenue to gain management's attention. No executive wants to be warned about a serious problem without also hearing a viable solution. As the FSD responsible for the facility, it is important to know how hazardous materials must be incorporated into your fire operation plan.

In an RAA, you must investigate all possible problems and solutions concerning the storage and transportation of hazardous materials. To accomplish this, review and analyze your facility's situation, and consult with other department managers to gather their input. List those materials stored in the facility and the effects they have on the safety of the persons exposed to them. While the safety of these particular individuals is the immediate concern, inform administration of how hazardous materials affect all persons working at or visiting the facility.

Administration must be prepared to work with the Occupational Safety and Health Administration (OSHA) and local fire departments when dealing with hazardous materials, and still keep the facility's operating costs to a minimum. *The Factory Mutual System Approval Guide* notes "that meeting the standards of the Occupational Safety and Health Act and environmental protection laws requires substantial capital outlays, yet contributes nothing to the capacity of production."[82]

It is your responsibility to convince management to think in terms of safety, not economics. Allocating funds to deal with emergencies is standard business practice. Yet is your facility willing to absorb the cost of complying with OSHA standards and training your staff to handle a crisis involving hazardous materials?

Citing responses to a survey commissioned by the NFPA, *The Factory Mutual Systems Approval Guide* states "that managers of many of the nation's 500 largest industrial organizations do not assign priority to the critical area of loss prevention engineering and the insurance risk problem. In that leadership group, it is imperative that every member recognize the problem."[82] While economic factors are important, the protection of life and property should be the priority at all times.

OSHA has established hazardous waste and emergency response regulations. These regulations safeguard workers responding to hazardous materials emergencies. Provisions of compliance entail a drafted safety and health program that provides emergency response procedures. Although environmental compliance has been established, many organizations are slow in complying with these regulations. It is imperative you do not take a wait-and-see attitude regarding compliance. According to Brent Bassett, executive officer of BDN Industrial Hygiene Consultants Incorporated, in Portage, Michigan, consulting services are available to individuals who are not familiar with hazardous materials. Consulting services will identify needs and justify appropriate training in specific facilities. However, Bassett indicates that some facilities will require minimal training if the city in which they work has good emergency response procedures. If emergency response is inadequate, additional training will be required.

The Superfund Amendments Reauthorization Act of 1986 (SARA III) ensures that individuals can find out the chemicals that are present in their workplaces as well as in their communities. *Title III, "Emergency Planning and Community Right-to-Know,"* affords the creation and maintenance of an emergency planning and accounting arrangement for all facilities that use chemicals, including industrial companies, hospitals, and learning institutions.

Since the law's provisions have been enacted, company executives and safety, environmental, and fire safety professionals have prepared to deal with the information required by government agencies and communities. Information is gathered and maintained regarding chemical storage and the handling and disposal of these chemicals. SARA III provides the means of increasing the general public's awareness of products, safety measures, and operational standards. It also requires that companies plan for a crisis and then practice implementing the plan.

The Material Safety Data Sheet (MSDS) *Pocket Dictionary* is a reference booklet that all security/FSD's should have. The MSDS *Pocket Dictionary* is published by Genium Publishing Corporation, of Schenectady, New York, and is written with the objective to concisely inform the reader of the hazards of the materials with which one works or is exposed. This booklet or pocket dictionary is written as a guide for persons who routinely work with hazardous materials.

The MSDS *Pocket Dictionary* explains why the format of the MSDS must comply with specific legal requirements. It also lists and examines the nine relevant sections detailed on any MSDS—material identification; ingredients and hazards; physical data; fire and explosion data; reactivity data; health hazard information; spill, leak, and disposal procedures; special protection information; and special precautions and comments.

The necessity, not only of reading, but of also comprehending an MSDS to safeguard life and property in an industrial setting is understandable. The MSDS *Pocket Dictionary* plays a critical role in aiding safety and security personnel, management, and other employees who directly handle hazardous material.

A dictionary of terms comprises the majority of the text and includes terms and abbreviations pertaining to an MSDS. Certain acts and laws are also explained in the dictionary, complete with addresses and telephone numbers for the reader, if questions should arise.

PERSONAL PROTECTIVE EQUIPMENT

According to the United States Department of Labor Occupational Safety and Health Administration there are specific, established requirements pertaining to personal protective equipment.

Protective equipment should be supplied by your organization when you have chemicals or hazardous material stored or in transit to your facility. It is necessary to have protective equipment for the eyes, face, head, and extremities; and to provide protective clothing, and respiratory devices.

If the employee owns the equipment, the employer shall be responsible for ensuring the equipment's adequacy, maintenance, and sanitation. All personal protective equipment should be designed and constructed according to the specifications of OSHA. All equipment designed and constructed to protect the face, head, and eyes must be in accordance with the American National Standard for Occupational and Educational Eye and Face Protection. Be sure that the equipment your facility uses to protect employees conforms to the standards of the various agencies who have jurisdiction in these areas.

Fire fighters are usually well equipped with protective clothing by their department. They are trained to never enter a burning building unless wearing full protective clothing, including fire helmets. It is imperative that you advise your administration to adjust budgets because special equipment is needed for personnel who combat fires and handle hazardous materials. Your staff should be as well protected as the municipality's fire department when fighting a fire or responding to a chemical spill. Never underestimate what your administration will do for your department once it understands and recognizes the hazards associated with the job.

HALOGENATED EXTINGUISHING AGENTS

Water is the most conventional extinguishing agent; it cools the burning elements and smothers the fire. For special situations, however, other extinguishing agents are more effective. Fire-fighting foam is used predominately to combat fires involving flammable liquids. Carbon dioxide suppresses fires involving gas, flammable liquids, electrical equipment, and common combustible elements, such as paper or wood. However, to extinguish computers and other expensive electrical office equipment, halon is used. Halon produces a dense, nonflammable, nonconducting vapor that leaves inconsiderable residue after it is applied. Halon is also used to suppress fires on airplanes.

Because you may have offices, electrical rooms, and computer rooms that are already equipped with a sprinkler system, it is imperative you understand when you would want to consider using halon, instead of water. While water will suppress and contain a fire, it may also cause irreversible damage to computers and electrical equipment.

> Halogenated extinguishing agents are hydrocarbons in which one or more hydrogen atoms have been replaced by halogen atoms. The common halogen elements from the halogen series are fluorine, chlorine, bromine, and iodine. The hydrocarbons from which the halogenated extinguishing agents are derived are highly flammable gases and, in many cases, the substitution of halogen atoms confers not only nonflammability but flame extinguishment properties to the resulting compounds.[84]

Halogenated agents applies to both fractional and total substituted hydrocarbons. Not all fractional halogenated compounds are nonflammable. For example, methyl chloride is flammable, whereas methyl bromide is nonflammable under normal conditions. You should know about halogenated agents and their effects in fighting fire. You must be aware if the facility uses any halogenated agents and where they are stored. When fighting a fire, your staff as well as the fire department must be made aware that these agents are being used. There can be toxic and irritant effects from short-term exposure to halogenated agents.

"Consideration of the life safety of halogenated agents does not stop with the effects of the agents themselves, but must also be given to the effects of breakdown products which may have relatively higher toxicity. Decomposition of halogenated agents takes place on exposure to flame, or to surface temperatures above approximately 900°F."[85]

Halogenated extinguishants are expensive and this is why many facilities will only put halogenated agents in specific areas of their facility. When you deal with these agents, it is imperative that you understand what halon agent numbers represent. For example, what does halon 1011, or halon 1301, or halon 1211 represent? "In a halon number, the first digit on the left indicates the number of carbon atoms in the basic molecule; the second, the number of flourine atoms; the third, fourth, and fifth digits indicate chlorine, bromine, and iodine atoms. Zeros are used between other numbers but are not used as last or first digits (carbon will always be present). Bromochlorodifluoromethane (CBrCIF2) is therefore designated Halon 1211. Halon numbers are used only with halogenated hydrocarbons."[86]

PRECARIOUS FIRE SUPPRESSANTS

Experience has taught us that we learn from our mistakes. Sandra Carey once said, "Never mistake knowledge for wisdom. One helps you make a living; the other helps you make a life." In fire safety, one mistake can cost many lives. However, the knowledge and wisdom we have acquired through the years in the field of fire safety continues to save lives. Education, training, and experience go hand in hand with fire safety. Like any other field of science, fire science continues to grow by leaps and bounds. Be aware that using certain fire suppressants under the wrong conditions can be deadly. As previously indicated, water is the most widely used suppressant in fire fighting. However, water should not be used on very hot or burning magnesium. This can cause a violent reaction to occur. The water disintegrates and discharges hydrogen that will explosively ignite as it disperses into air. There are other water-reactive chemicals that can have the same effect. Be aware of them:

acetyl bromide	potassium
acetyl chloride	potassium peroxide
aluminum trialkyls	potassium hydroxide (solid)
calcium	rubidium
calcium oxide	sodium
diborane	sodium amide
dimethyl sulfate	sodium hydride
lithium	sodium hydroxide (solid)
phosphorus oxychloride	sodium peroxide
phosphorus trichloride	sulfur chloride

Your MSDS *Pocket Dictionary* better explains the composition of these chemicals as well as the proper storing and handling of them. In the event of a fire, be careful not to use water on these chemicals.

Prepare a chart of the water-reactive chemicals stored at your facility. List where they are stored, how they are stored, and to which departments they will be shipped. Meticulous record keeping and proper storage of these chemicals is necessary to avoid explosions, spills, or leakage.

Many times, fire departments use streams of water to fight certain hazardous material fires. The primary effect of a solid stream of water concentrated on a specific area of a fire is the cooling of the ignited fuel below the fire. Keep in mind that this does not provide a blanketing effect. It is advantageous to fight fires from great distances using solid streams of water. When a fire exists, either the flames or the combustibles may be cooled to a temperature below the point required for the fire to continue. Of the two, cooling the combustible is much more effective. This can be done by a coolant absorbing sensible heat when its temperature is raised; or, more effectively, by absorbing heat required for vaporization of a coolant. Blowing a flame away will keep the heat of reaction from reaching the surface of a flammable liquid or solid. A cold quenching device or surface may absorb enough heat to prevent propagation either through a flammable gas mixture, or back to a solid liquid.[87]

STORAGE REQUIREMENTS

There are four classifications listed by the NPFA regarding general storage requirements. Become acquainted with these classifications.

Class 1

Class 1 is classified as stock that is essentially noncombustible products stored on combustible pallets. This stock can also be stored in corrugated boxes with or without individual thickness dividers or wrapped in ordinary wrapping paper and placed on or off pallets.

Class 2

Class 2 is classified as stock that is defined as Class 1 products stored in slatted boarded crates, solid wooden containers, multiple density paper-board containers, or equivalent combustible packaging fabric and placed on or off pallets.

Class 3

Class 3 is classified as stock that is defined as wood, paper, or natural tendril fabric or products stored on or off pallets. This stock may contain a limited amount of plastic. Any stock or merchandise that contains plastic is given a Class 3 classification.

Class 4

Class 4 is classified as stock that is defined as Class 1, 2, and 3, and is composed of a considerable quantity of plastic stored in plain, furrowed cartons or Class 1, 2, and 3, stock that is stored in plain furrowed boxes containing plastic packing and stored on or off pallets. Class 4 stock pertains only to non-capsulated products.

Storage occupancies include facilities or structures utilized for the primary purpose of storing or sheltering merchandise, products, vehicles, or animals. The NFPA lists several storage occupancies that are affected by the Life Safety Code. Do not mistake the storage of stock and merchandise for the industrial occupancies that make these products. These are two distinct classified occupancies. The NFPA lists the following storage occupancies:

warehouses	cold storage
freight terminals	truck and marine terminals
bulk oil storage	parking garages
hangars	grain elevators
barns	stables

Industrial occupancies are facilities that make and produce all kinds of merchandise devoted to operations such as processing, assembling, mixing, packaging, finishing, and decorating. Industrial occupancies can be considered high-hazard occupancies. Many times, gasoline and other flammable liquids are handled, used, produced, and stored at these facilities.

The following industrial occupancies are listed by the NFPA:

factories of all kinds	laboratories
dry cleaning plants	power plants
pumping stations	smokehouses
laundries	creameries
gas plants	refineries
sawmills	college and university noninstructional laboratories

The NFPA has revised the requirements for maximum travel distance to exits in industrial occupancies. "In most cases the requirements for maximum travel distance to exits will be the determining factor rather than numbers of occupants, as exits provided to satisfy travel distance requirements will be sufficient to provide exit capacity for all occupants, except cases of unusual arrangement of buildings or high occupant load of a general manufacturing occupancy."[88]

Another factor of which the FSD should be aware is the fact that these products are stored under conditions that may involve the release of flammable vapors.

Who determines the hazard content of your facility? According to the NFPA, the hazard content of your facility shall be determined by the local authority having domain regarding the classification of your facility. The contents and the processes conducted in the facility has a direct bearing on hazard content.

The NFPA has determined that the hazard content of any facility shall be classified as low, ordinary, or high in accordance with the Life Safety Code's section on "Hazard of Contents."

"*Low hazard* contents shall be classified as those of such low combustibility that no self-propagating fire therein can occur and that, consequently, the only probable danger requiring the use of emergency exits will be from panic, fumes, smoke, or fire from some external source.

"*Ordinary hazard* contents shall be classified as those which are liable to burn with moderate rapidity or to give off a considerable volume of smoke, but from which neither poisonous fumes nor explosions are to be feared in case of fire."

"*High hazard* contents shall be classified as those which are liable to burn with extreme rapidity or from which poisonous fumes or explosions are to be feared in the event of a fire."[89]

It is crucial for you to interface with the manager of the industrial facility to ensure compliance with the provisions set forth by the municipality. Proper storage arrangements are vital to the facility as well as the FSD, who is trying to maintain a fire-safe environment.

If your storage or industrial facility is equipped with a sprinkler system, it is recommended that the sprinkler be designed to provide adequate water density to control a fire. "In unsprinklered buildings, good practice is to provide minimum separation of 8 ft between piles, with pile sizes, excluding surrounding clear spaces, limited as follows:

Noncombustible	**No Limit**
Class I & II	15,000 sq.ft.
Class III	10,000 sq.ft.
Class IV	5,000 sq.ft."[90]

BOILING LIQUID-EXPANDING VAPOR EXPLOSION (BLEVE)

If your facility transports or receives hazardous materials, you should understand how these materials are transported and how they should be stored. Most chemicals are transported via truck or railroad car. The chemicals are stored in liquid form within the containers of the truck or railroad car. Studies and research have aided fire engineers in designing suitable containers and appropriate ways of transporting hazardous materials. "Containers of compressed or liquified gases can represent high levels of potential energy release due to concentration of matter by compression or liquification. Container failure releases this energy—often extremely rapidly and violently—with simultaneous release of gas to the surroundings and propulsion of the container or container pieces. Compressed gas container failures are distinguished more by the flying missile hazard than by the results of the gas release because they contain lesser quantities of gas. Liquified gas container failures can release larger quantities of gas."[91]

Most BLEVEs occur in the vapor space and are identified by the metal expanding and thinning. This expansion gives the impression of a longitudinal tear that gets progressively larger and larger until the critical span is reached. When this point is reached, a violent chemical reaction occurs and a tremendous explosion results.

Begin to formulate strategic plans to see that the containers transporting the hazardous materials are safe. In North America, there are two types of gas containers—*cylinders* and *tanks.*

Gas cylinders are fabricated in accordance with the regulations and specifications of the Department of Transportation (DOT). The DOT comes under the jurisdiction of the United States Department of Transportation and the Canadian Transportation Commission. "The regulations cover the service pressure the cylinder must be designed for, the gas group or gases that it can contain, safety devices, and requirements for in-service (transportation) testing and requalification. The specifications cover such criteria as metal composition and physical testing, wall thickness, joining methods, nature of openings in the container, heat treatments, proof testing, and marking."[92]

Gas tanks are fabricated so that they comply with Section VIII (Unfired Pressure Vessels) of the Boiler and Pressure Vessel Code promoted by the American Society of Mechanical Engineers (ASME) or tank fabrication standards of the American Petroleum Institute (API).

"The DOT and Canadian Transportation Condition (CTC) Regulations, in themselves, apply only to cylinders and tanks in transportation in interstate or interprovincial commerce. However, many consensus codes and standards extend these regulations to transportation in intrastate commerce and also extend applicable criteria to use and storage at consumer sites."[93]

You cannot eliminate all hazards from a facility, but you can provide safeguards by checking and safeguarding trucks that transport hazardous material to the facility. When trucks are en route to your facility, request that a check of the truck be made by the driver and a member of the fire safety department. Check for leaks, spills, loose valves, and odors. If any of these conditions are present, do not allow the truck to proceed until the source of the concern is discovered.

Plan ahead and choose an isolated area of your facility that a truck may be stored should it arrive in an unsafe condition. If you cannot place the truck in an isolated area, remove all other vehicles and products from the area in which the truck is located. If the truck is emitting vapors, keep all persons away. Notify your local fire department and then notify CHEMTREC at (800) 424–9300. The operator will ask you if you have a chemical spill; give the operator as much information as possible. He will advise you what to do and what emergency action to take. Some suggestions he may give you include

notify your local fire department

keep the area free from pedestrians

stay upwind; stay out of low areas

isolate the hazard area and deny entry

wear a self-contained breathing apparatus and full body protective clothing

await the arrival of the fire department for further instructions

If the truck is on fire, immediately notify the fire department, explain that it is a chemical fire with toxic material, and evacuate the area. Try to evacuate persons within a ½ mile radius. BLEVE explosions have been known to launch projectiles up to ½ mile away.

If your facility has tanks, determine their type and when they were last cleaned, checked, and inspected. "Tanks designed for installation in fire resistive enclosures within buildings have the same metal thickness and design features as required for all tanks. Fuel oil tanks designed for use without being enclosed in a fire resistive, cut off room inside buildings are normally restricted to less than 660 gallon capacity. Although the metal thickness for the tanks are designed for differing situations they are approximately the same, the location of the openings for pipe connections differ. Pipe connections for underground tanks and enclosed fuel oil tanks inside buildings are in the top only, whereas unenclosed tanks are provided with bottom outlets for gravity feed piping to such installations as oil burning equipment."[94]

Storage tanks within structures will vary depending on the class of the liquid and the occupancy of the building. The NFPA lists specific requirements for the installation of tanks within buildings. The NFPA's Flammable and Combustible Liquids Code specification is the source to which you should refer if you have any questions.

SUMMARY

The field of fire science continues to grow by leaps and bounds. Scientists and mechanical, electrical, and architectural engineers have made tremendous advancements. Due to their efforts, the planning for education and training, preventing, combatting, and containing fires has been greatly improved. The myriad fire safety and life safety codes that have been established over the past three decades can be largely attributed to professionals' work in the fire safety field.

With every fire, there are new frontiers to explore. *How* did the fire start, *where* did the fire start, and *why* did the fire start are questions investigators ask after a fire has been extinguished. Investigators try and find answers that will assist others in the field of fire science. One individual's discovery may unlock answers we have sought for years.

Through years of research and study, we have learned about the cooling effects of water and the effects of halogenated agents used to combat and suppress fires. We have also learned that water should not be used on certain chemicals or metals. Water, sprayed on the wrong chemical or metal, will cause a violent reaction to take place.

The NFPA has regulations for storage of hazardous material in industrial occupancies. Research has taught us that it is beneficial to separate buildings by classifications. These classifications allow for the safe storage and production of combustible and noncombustible stock.

The hazard content of your facility is vital to your fire operations plan. Is your facility classified as having a low, ordinary, or high hazard content? Check with your municipality to determine your facility's classification.

If your facility stores or transports hazardous materials, be aware of what to do in the event of a spill, leak, or fire involving these materials. How you react to a chemical spill is crucial. The handling and transportation of these materials should be part of your fire operation plan in a section for hazardous materials.

The storage tanks within your facility are yet another cause for concern. Usually, the building manager or superintendent is responsible for the care and maintenance of these tanks. However, should a spill, leak, or explosion occur, it is your responsibility to evacuate the facility and deal with the hazard.

Dealing with Adversity

Designing and implementing a flawless security and fire safety system is virtually impossible. Just when you think you have incorporated every aspect into your plan, a piece of the puzzle won't fit. What should you do? Deal with adversity by incorporating and changing your system as needed. True professionals always think of solutions to problems they encounter.

The previous chapters have discussed various ways to design and implement fire safety plans and procedures, and how to choose fire safety officers. This book contains information that novice FSDs may not understand at first. However, after a few months at their jobs, they will better understand this material.

Let's take, for example, the director of security who has just been appointed FSD of a residential complex. There are three buildings, 17 stories each, including a basement. There are a total of nine apartments per floor, with the exception of the basement and first floor. The first floor of each building has four apartments. The basement contains a laundry room, the superintendent's apartment, and a maintenance shop. There is an outdoor parking area for 60 vehicles and one underground parking garage for 200 vehicles. The security office is located in the basement of the middle building. The management and renting office is located on the first floor of the first building. You are responsible for 504 guard hours. What do you do regarding fire safety and where do you begin?

Begin by conducting a property fire prevention survey of your facility. Ask management if the previous director implemented a fire operation plan. If the previous FSD did not have a plan, begin formulating one. If the previous FSD did have a plan, review it, and work from there. Compile information regarding the topics listed in this chapter. Add or subtract additional categories of information depending on the classification of your facility.

BUILDING STRUCTURE CLASSIFICATION

You know there are three buildings, each 17 stories tall. Determine how many apartments are included throughout all three buildings and how many vacant apartments are present throughout the complex. Inquire when the buildings were constructed and determine their construction. Ascertain if any sprinkler systems are in the buildings. Inquire if there are any places of assembly used by the residents for meetings. Count the number of exits and entrances into each building. How many stairwells are within each building? Are there standpipes within the stairwells?

Up until a few years ago, this information would have been sufficient. However, an experienced FSD will also inquire the names and apartment numbers of residents who are physically handicapped. Does the property have ramps for the handicapped? If not, would management allow the construction of ramps for the handicapped? Does the property have any residents who are blind? If so, what precautions has management taken to ensure their safety?

Does your property have any persons who are hearing impaired? If so, smoke detectors are of no use to these residents. What would you recommend? The hearing impaired have begun using visual and vibratile alarm signals as waking devices. Smoke alarms, which have a visual effect rather than an audible effect, are being used for the hearing impaired. Strobe lights and incandescent lights within the smoke detector give off bursts of light rather than an audible signal.

FIRE/BUILDING SURVEY

Ascertain the last time the property was surveyed or inspected by the fire or building departments. Inquire if the property has any outstanding fire or building violations. If it does, correct the violations within a reasonable amount of time, not to exceed 30 days.

Implement your fire safety audit sheet. Advise your staff that, effective immediately, fire safety will become a primary consideration of the security department. Review these audit sheets and take immediate action if any deficiencies are found. Always establish and maintain a rapport with your maintenance department as you will be asking them to correct the majority of the deficiencies.

FIRE OPERATION PLAN

If your predecessor did not have an approved fire operation plan, develop one. If you inherit a fire operation plan, be sure it is updated and meets the requirements of the local municipality.

Review the plan with members of the maintenance department. Seek their input and guidance. Remember, as the new kid on the block, you need all the help you can get.

Write the fire operation plan to suit the type of facility and occupancy for which you are responsible. The fire operation plan for this residential property must take into account handicapped individuals, senior citizens, and children.

Do you have sufficient and properly trained personnel available for all fireguard patrols to satisfy what is required by your municipality? If not, select and train qualified candidates to be part of your fire brigade.

Know where you are going to situate your nerve center in the event of a fire emergency. Know the location of the nearest fire alarm pull box and establish a location at which you will meet the arriving fire department.

DETECTION, ALARM, AND
COMMUNICATIONS SYSTEMS (DAC)

Inquire about the facility's DAC. What type of detection systems are throughout the facility? Are these systems tied into a central station or are they tied into the fire command center? If there is a fire detection or alarm system, inquire when it was last tested. If the previous FSD kept accurate records, all this information should be properly recorded.

In reviewing DAC, inquire if there are any special features regarding these systems. If there are special features, they should be recorded and posted by the fire command center.

Inquire if security personnel are familiar with DAC. If they are not properly trained, make this one of your primary concerns.

What if the FSD discovers that all three buildings do not have DAC tied into a central station or the nerve center? A separate fire operation plan must be formulated for each of the three buildings.

Most residential buildings do not have DAC throughout the facility. How, then, will a fire or smoke condition be reported to the security nerve center? In most cases, fire or smoke conditions are reported via telephone by a neighbor or a pedestrian. Your fireguard should be specifically assigned to detect a fire or smoke condition.

SECURITY/FIRE PATROL GUARDS

Most residential properties are more concerned with security than fire safety. It is your responsibility to justify your budget and request that one fireguard per shift conduct fire watch tours. The security/fire patrol guard will serve as both a patrol officer looking for and deterring criminal activity and a conductor of fire watch tours. The security/fire patrol guards should patrol the facility with a Fire Safety Audit Sheet. Inform management that these tours are vital for a fire-safe complex and will ensure the safety of all residents, employees, and visitors.

Ensure that these fire patrols are effective. The members of the fire patrol must be skilled in detecting fire hazards and taking corrective action regarding these hazards. All fire systems, as well as standpipes and sprinkler systems, must be checked during a guard's tour. The guard will report in, via walkie-talkie, every ½ hour and will provide a security and fire safety report. Obviously, if an incident occurs, the guard should report it immediately.

CONDUCTING FIRE DRILLS

Your biggest obstacle regarding the residential complex is conducting effective fire drills. These drills must be conducted on a monthly basis and participation is vital. However, tenants rarely participate in fire drills, especially if the drills are voluntary. How, then, does the FSD run an effective fire drill without tenant participation?

Ask the residents if they wish to participate in the fire drills. If they won't, conduct the fire drills without them. Ask to speak at the monthly board or tenant meetings. You may gain some support and cooperation, and you can include those residents who wish to participate. However, mandatory participation is not necessary or required by law.

Advise tenants that, as the security/FSD, you take fire safety very seriously. Inform them that fire drills will be conducted on a monthly basis. Prepare written instructions for all tenants regarding what to do in the event of a fire or smoke condition. Assist tenants in identifying stairwell locations. Post *You Are Here* signs on each floor. These signs must advise the occupant of the route to take in the event of a fire. By making residents aware of fire safety problems, and by telling them that it is for their own safety and protection, you begin to gain their support and confidence.

SUMMARY

Use these five criteria as your starting point—address building structure classification, fire operation plan, DAC, security/fire patrols, and fire drills. Your overall fire operation plan may take months, sometimes even years, to formulate. Set deadlines to help you complete portions of your fire operation plan. When any type of adversity arises, meet it head on and deal with it. Use research books, ask questions, consult with professionals. Problems do not go away. The more quickly you and your staff address the adversity, the faster it will be resolved.

Not every security director or FSD will deal with a crisis or adversity in the same way. Police science is vastly different from fire science; security is vastly different from fire safety.

In fire safety and fire prevention, you face challenges daily. A potential crisis can occur simply by receiving an unexpected shipment of hazardous chemicals.

It is imperative that all persons involved in fire safety and fire prevention know their facility, their employees, and read as much as possible regarding fire safety. The FSD who does not research codes, mandates, and regulations will always deal with more adversity than an FSD who does his research. Codes and mandates become outdated, and local municipalities can change a local law without you realizing it.

Be flexible when dealing with adversity. A colleague of mine once criticized me for stating that sand buckets could be used in garages in lieu of fire extinguishers. He advised me that this method of fire fighting was impractical in today's society. I disagreed. I also informed him that there was no budget to purchase the required number of fire extinguishers for the garage. I checked with the local fire department and they said sand buckets would do just fine. They were old-fashioned, but fine. I placed the sand buckets in the garage, kept the co-op board happy, and stayed within my fire safety budget.

Deal with adversity head-on. All professionals do. Take pride in analyzing and overcoming adversity. Seek input from those individuals whom you respect and who are knowledgeable in the fire safety field.

12

The Overall Importance of Standpipes and Hoses

A standpipe system is a pipe structure designed to transfer water, vertically to upper floors of high-rise buildings or horizontally throughout massive areas of buildings, for fire fighting purposes. The concept of the standpipe system is to provide a means of supplying water to a fire or smoke condition without resorting to extended fire hoses. Fire fighters or trained fire brigade members who operate these hose lines are more effective than firefighters using extended fire hoses in fighting or containing a fire or smoke condition for that particular building.

A standpipe system should have outlets on each floor and an adequate number of risers, depending on the area of the facility. A good rule of thumb is that each linen hose should have a maximum length of 125 feet. In this way, every point on a floor will be within 20 feet of a nozzle. It is vital that you or a member of the fire brigade inspect these hoses monthly for proper placement and storage, and for evidence of tears or decay.

Unlined linen hose is a material hose fabricated of closely woven linen. Due to the typical attributes of linen, the threads will swell when wet and will close the minute spaces between the threads, thereby making the hose watertight.

Only trained professionals should be permitted to use the standpipe system. Trained fire brigade members or fire fighters are instructed in the installation of these hoses during a fire condition. If seepage occurs immediately after the water is turned on, it is an indication that the hose is inadequate and should be replaced.

After use, these hoses must be dried properly. Hoses tend to deteriorate quickly if not thoroughly dried immediately after use. Do not store these hoses outside (unless they are encased) and do not expose them to dampness.

The racks that are used to store linen hoses are important to the effectiveness of the hoses. Racks are constructed to allow for the appropriate circulation of air about the hose and to protect it against dampness. (See Figure 12–1) Underwriters Laboratories specifies the type of hose racks that satisfy their standards.

167

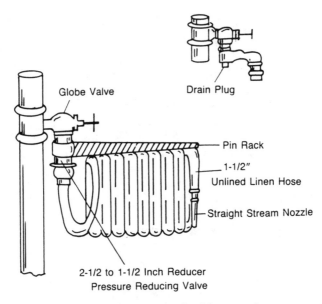

Globe Valve

Drain Plug

Pin Rack

1-1/2"
Unlined Linen Hose

Straight Stream Nozzle

2-1/2 to 1-1/2 Inch Reducer
Pressure Reducing Valve

Figure 12–1   A standard hose rack.

The linen hose should be unfolded and refolded at intervals to avoid permanent creases. Gaskets in couplings at the hose valves and the nozzle should be checked. It is crucial that you inspect the valves and nozzle when inspecting the hose.

Hose valves must be closed tightly to prevent the linen hose from becoming wet due to leakage. *Drip cocks* can be placed in the hose valve assembly to drain any water that may collect. Ask your local fire department or the company that installed the standpipe system about the use of drip cocks.

Incorporate pressure tests of standpipe systems and hoses in your fire operation plan. A pressure test uncovers weaknesses in portions of the hose and will determine if the hose has the strength to withstand a hydrostatic pressure of 25% in excess of the standard pressure on the standpipe.

Standpipe and hose systems should have adequate water pressure continuously available. When this is unrealistic, such as in an unheated structure, the standpipe should be systemized to admit water automatically via a dry pipe valve or other NFPA-approved device.

FOUR TYPES OF SYSTEMS

There are four standpipe systems recognized by the NFPA. If your facility utilizes a standpipe system, be sure you know its type.

I: The Wet Standpipe System

The wet standpipe system has an open supply valve and continuously sustained water pressure. According to experts, this system is more effective in fighting fires than any of the three other systems.

II: The Dry Standpipe System

The dry standpipe system is designed and situated to allow water to flow through it via manual operation or by specified and sanctioned remote control apparatuses located at each hose station.

III: The Dry Standpipe System in an Unheated Facility

The dry standpipe system in an unheated facility should be capable of allowing water to flow automatically via a dry pipe valve or any approved apparatus. The lack of air in the system at the time of use causes a delay in the water reaching the smoke or fire area.

IV: The Dry Standpipe System Without a Fixed Water Supply

In high-rise buildings, the dry standpipe system without a fixed water supply is used. This system decreases the time needed for fire departments to install hose lines on upper floors of high-rise buildings. During construction, this system may be employed in lieu of the wet standpipe system in unheated facilities. The wet standpipe system would not be used in an unheated facility because the water in the pipes would freeze.

SYSTEM CLASSIFICATION

There are three classifications regarding standpipe and hose systems. Some jurisdictions have eliminated the criteria for occupant-use hose systems in facilities that are entirely protected by automatic sprinklers. Check with your municipality to determine which mandates apply to your facility.

"Class I systems (2-1/2 in. hose connections) are provided for use by fire departments and those trained in handling heavy fire streams. In nonsprinklered high-rise buildings beyond the reach of fire department ladders, Class I systems can provide water supply for the primary means of fire fighting, i.e., manual attack on fire.

Class II systems (1-1/2 in. hose lines) are provided for use by the building occupants until the fire department arrives. The hose is connected to 3/8 or 1/2 in. open nozzle or combination spray/straight stream nozzle with shutoff valves. Shutoff or spray nozzles are seldom provided unless the occupancy is one where hand hoses would be used frequently. Normally, the hose is kept attached to the shutoff valves at the outlets. Where the occupant-use hose streams can be properly supplied by connections to the risers of wet-pipe automatic sprinkler systems, separate standpipes for these small streams are not required.

Class III systems are provided for use by either fire departments and those trained in handling heavy hose streams, or by the building occupants. Because of the multiple use, this type of system is provided with both 2-1/2 in. hose connections (for use by fire departments or those trained in handling heavy hose streams) and 1-1/2 in. hose connections (for use by building occupants). One method for accommodating this multiple use is by means of a 2-1/2 in. hose valve with an easily removable 2-1/2 × 1-1/2 in. adapter, permanently attached to the standpipe."[95]

Most security directors or novice FSDs read the previous paragraphs and cringe. Do not be intimidated by numbers or language that is foreign to you! Whenever you survey a building with standpipes, sprinkler systems, or sophisticated fire suppression systems, automatically ask to speak with the building engineer or building operations manager. The building engineer will know more about the design, components, and operation of the building than anyone else.

Over the years, I have developed working relationships with many building engineers. Building engineers have always acted professionally and gladly furnished information and advice regarding a building's fire systems. One friend in particular, Robert Sweeney, a building engineer in New York City, has proven to be a valued friend and a knowledgeable person in the field of fire systems. Whenever a question arises regarding building operations on fire systems, I phone him and seek his advice. If he does not know the answer, he will contact others in the field who will find the answer. It is imperative that you build a network and make acquaintances in order to better understand the intriguing field of fire safety.

Many times, arguments develop among colleagues of mine regarding fire department connections. It is imperative that you fully understand what the NFPA means regarding approved fire department connections. "An approved fire department connection is to be provided for each Class I or Class III standpipe system. For high-rise buildings having two or more zones, at least one fire department connection is needed for each zone. Fire department connections are on the street side of buildings near fire hydrants for easy connection to fire department pumpers, and are to be marked either **STANDPIPE** or **STANDPIPE AND AUTO. SPKR.**, depending on the service."[96]

You should also be familiar with approved gauges as well as automatic alarm and supervisory equipment. "An approved 3-1/2 in. dial spring pressure gauge is connected to each discharge pipe from fire pumps, to each supply connection from the public waterworks, at each pressure tank as well as at the air pump supplying the pressure tank, and at the top of each standpipe."[97] Alarm detection systems such as water flow devices and tamper switches can be interfaced with the standpipe system to send a signal locally or to a central station when the standpipe is activated.

The NFPA Standpipe and Hose Systems Standard limits zone height to 275 feet. However, the height may be elevated to 400 feet when a pressure-decreasing mechanism, controlling nozzle pressure during both flow and no-flow situations, is installed at each opening.

The NFPA designed and simplified zoning. "The best practice is to divide the tall building into pressure zones. About twelve stories in a zone results in water pressure being about normal. This simplifies fire protection measures, as excessive pressure for hose lines does not have to be compensated for by pressure reducing devices and other complications. Water storage for fire protection should be calculated for each pressure zone much as one calculates storage areas of buildings on hills or other elevations supplied by municipal systems. Tanks providing storage in each pressure zone may be filled from the piping supplying water for other purposes in the building. Each pressure zone should have a gravity tank and a fire pump, the latter taking suction from the gravity tank in the next lower zone. Each zone should have its own fire department connection."[98]

One factor that drives building engineers crazy is cold temperatures. When sprinklers or standpipes are exposed to temperatures of 32°F or lower, they can freeze. This can be a building engineer's or an FSD's nightmare. Through research and design, individuals in the field of fire science have begun to mitigate this problem. Dry compressed nitrogen in cylinders, in substitution of air, decreases the accumulation of moisture in the system. Propylene glycol or other suitable chemicals may be substituted for the priming water. Adding a small dose of mineral oil to the surface of the priming water has been known to prevent evaporation. Propyleneglycol, a nonvolatile priming liquid, can be used for the same purpose.

If your facility has a combined sprinkler and standpipe system, the sprinkler risers can be utilized for feeding the sprinkler system and the hose outlets. The outlets are 2-1/2 inches. If the facility is entirely sprinkled, 1-1/2 in. hoses for inhabitant use can be excluded.

In most cities, combination sprinkler-standpipe systems have yellow-painted Siamese caps. The purpose of the standpipe Siamese cap is to increase the volume and pressure of water being supplied to the system. In most cities, the standpipe systems will have red-painted Siamese caps.

Tests of the standpipe system will be conducted by qualified individuals who are certified and experienced in testing standpipes. Standpipe tests will demonstrate, to the satisfaction of the local municipality responsible for testing the system, that the system will sustain a hydrostatic pressure sufficient to produce a pressure of at least 100 pounds per square inch (PSI) at the top-story hose outlet and at least 100 PSI at the Siamese hose connection. Maintain accurate records regarding the testing of the standpipe and sprinkler systems. A hydrostatic pressure test of the system should be performed yearly.

SUMMARY

You must become familiar with standpipe systems. Standpipe systems are piping structures designed to transfer water to upper floors of high-rise buildings. The primary reason standpipe systems are most effective in combating fires or smoke conditions is that they are able to supply water without the fire department resorting to extended fire hoses.

There are four types of standpipe systems and each is vastly different from the other—the wet standpipe system, dry standpipe system, dry standpipe system in an unheated facility, and dry standpipe system without a fixed water supply. These systems are used in various building classifications.

There are three classifications for standpipe systems—Class I, II, and III. All three classifications define the type of system as well as how, where, and who may use them. It is important that you become familiar with these classifications.

Building engineers and FSDs both share a common goal—the protection of life and property. The building engineer knows all the building's operating systems as well as the fire safety systems. It is imperative that you and your staff build a rapport with the building engineer.

Know how the NFPA defines zoning and how it applies to your facility. The NFPA divides tall structures into pressure zones. Their rule of thumb is that about every twelve stories should have normal water pressure. Checking and inspecting your standpipe system is vital to your fire operation plan. Maintenance of these systems will only enhance their working effectiveness during a fire condition.

13

Evaluating Your Facility After a Fire

Your fire operation plan was in place, your DAC systems were operational, and your fire brigade acted effectively and efficiently. Why, then, did a fire occur and why did it cause extensive property damage?

The answers are not easy to find. A total evaluation of your facility after a fire has occurred is warranted in order to determine how the fire operation plan can be improved. If the damage was limited to property damage, with no loss of life, you should be grateful. However, your work is not completed after the fire is out. You, working with the fire department and building department, must begin to determine the following:

How did the fire begin?

Where did the fire originate?

Why did the fire originate?

Why did the fire spread?

Was arson involved?

Were DAC systems operational?

Was the fire operation plan effective?

Did members of the fire brigade do an effective job?

Was the fire department notified in sufficient time?

When will the facility be totally operational?

Answering these questions will take time. Submit a written report to administration when you have your answers. Advise administration to notify the insurance company so they may begin their investigation as well.

NFPA DEFINITIONS

The NFPA defines terms related to injuries and deaths associated with a fire. The NFPA Committee on Fire Reporting developed and instituted the following definition:

Fire Casualty: An individual who receives an injury or dies resulting from a fire. (The basis of fire casualties are categorized as direct or indirect. Direct fire casualties result in injuries or deaths due to fire. Indirect fire casualties result when injuries or deaths attributed to fire, but an additional catalyst is accredited as a predominant factor.)

Fire Injury: A fire injury is one that is received as an effect of a fire. This injury requires treatment by a medical practitioner for up to 1 year after the fire.

Fire Death: A fire death is a fatality caused by a fire up to 1 year after the fire.

More than One Cause of Death: This term is used when the fire is one of the causes of death. This term is further defined by the appropriate authorities in your municipality in conjunction with the Manual of the International Statistical Classification of Diseases, Injuries, and Causes of Death.

Indirect Causes of Fire Injuries: Injuries that are an indirect result of a fire. According to the NFPA, they can include

sickness acquired through exposure to weather during the time of a fire

injuries sustained during a fire that originated in a vehicle as a result of an automobile accident

overexertion while fleeing from or combating a fire

fire fighters who are injured when they respond to or return from a fire condition

fire fighters who are injured due to the violence of others when combating or containing a fire condition.

DISTURBING THE FIRE SCENE

After a fire has been extinguished in your facility and the fire chief has given his consent, keep all employees away from the fire scene. In most instances, the fire or police departments will cordon off the area with barriers or tape. Do not attempt to clean or visit the fire scene until the fire department, police department, and building department have concluded their investigations. Station a few security officers at the fire scene to keep spectators from touching or disturbing any evidence that may be beneficial to the investigation.

POST-FIRE ANALYSIS

When the fire department begins their investigation, request permission from the fire marshal to examine the fire scene along with the fire marshal. If a homicide occurred during the fire, this request may be denied. If the fire marshal *does* allow you to tour the fire scene, take notes and pictures of the scene. Many times, the fire marshal will provide you with valuable information regarding how the fire started, when it started, and how it spread.

A post-fire analysis will assist you, your department, and the fire department in evaluating the fire condition from its beginning to its conclusion. You will seek the answers to the questions who, what, when, where, why, and how. If the fire was small, such as a wastepaper basket fire, and the fire department was not notified, an investigation by you or a member of your staff is warranted. Obtain and record all the facts pertaining to all fires, big and small, and keep them filed in a confidential file. Pictures of all fires are not required, but it is a good practice to take pictures of all fire scenes and file them with your reports.

Four Goals

Your four primary considerations regarding the post-fire analysis are

1. Determine who, what, when, where, why, and how.
2. Determine if any criminal activity is associated with the fire.
3. Provide accurate information for the fire report.
4. Incorporate new procedures or revised procedures in your fire operation plan.

Procedures

An effective post-fire investigation will assist you in getting the facility operational again. It's crucial to have post-fire operational procedures. A subsection you may want to include in your fire operation plan is post-fire analysis and procedures. This subsection will be designed to answer the "what-if questions" regarding a fire condition. Questions that must be addressed are

After a fire occurs in the facility, is the company still capable of operating?

How soon after the fire will the company be at 100% operating capability?

Can the company temporarily relocate and continue operations or will it be forced to close down?

Will lack of work cause employees to be laid off?

Will company profits or sales be affected by the fire?

If the company is operational, are the DAC systems operational and capable of handling another fire condition?

Discuss these questions with administration and get their input and reactions prior to a fire condition.

FIRE INVESTIGATION AND REPORTING

As discussed earlier, other outside agencies or groups also have reasons to investigate fires. Insurance agencies will arrive at your facility within hours after a reported fire. The fire and insurance investigator will try and determine if the fire was caused for profit. If investigators determine this, the investigation will be lengthy. During any fire investigation, be as cooperative as possible with the investigators. After the investigators have determined the reason for the fire, their information will be used as input to underwriting and property protection programs.

"Special interest groups and manufacturers are interested in the performance of their product and will often investigate fires when there is a chance to gather information for product improvement or product safety, or to defend themselves in product liability cases.

Federal agencies with regulatory authority over certain hazards, conditions, or operations and code and standards development groups are interested in certain fires, as the information gained will provide input to the adequacy of the regulations or standards."[99]

The information the fire marshals or investigators uncover after a fire is recorded and stored for future use. This data contains vital information for the fire science field.

The NFPA realized that in order to evaluate and benefit from all fires a standard was needed in reporting fires. Therefore, the NFPA established the Technical Committee on Fire Reporting. This committee developed and instituted the Uniform Coding Standard. This standard classifies basic definitions and terminology used in fire reporting. The committee also codes the data it receives, making it possible to store and record all data in computer systems.

"The NFPA Uniform Coding Standard provides a common language for many data elements. It is recognized that every fire department will not want to collect every data element; likewise, there may be additional data elements that a fire department wishes to collect. When a fire department finds that it can use the data elements from the NFPA Uniform Coding Standard in its reporting system, it will also find that it is in a position to contribute data to larger data bases and to utilize data from these larger data bases in its municipal management."[100]

LEARNING FROM A FIRE

Let's assume that there was a fire at your facility. Nobody was injured, but there was approximately $1,000,000 in property damage. The fire marshal and the insurance investigators concluded their investigation and determined that the fire was caused by a faulty electrical wire in the basement. Where do you begin?

Gather the Facts

Your first obligation is to inform management verbally, then in writing, of the facts of the fire. Contact your fire communication officer and discuss, in detail, all the facts and events leading up to, during, and after the fire. Time sequences are important. Obtain the names of arriving fire captains, police officers, and ambulance attendants. Report the date and time of arrival and departure, and the names of fire marshals, insurance investigators, and other special interest groups who visit the fire scene. Describe what they discussed and indicate whether or not they took pictures of the fire scene. Describe the factors that contributed to the fire and caused the fire's ignition. Obtain this information from the fire marshal or insurance investigator.

When you have your facts, ascertain if the fire was foreseeable. Most electrical fires are difficult to foresee.

Evaluate DAC Systems

Evaluate the facility's DAC systems. Was the fire detected in its earliest stages? Did alarms sound or flash, thereby alerting employees and occupants to the fact that there was a fire or smoke condition? Did the fire communications system respond as it should? If the facility is sprinklered, did the system function properly? If the DAC systems operated effectively, state this. If, in your opinion, there was a problem, define it and recommend a solution to correct it.

Describe other fire systems that may be suitable for your facility. After a fire is a good time to get price quotes from companies to upgrade your present DAC systems. Inform administration that an improved system will benefit the facility and may also reduce insurance costs.

Check Code Compliance

Address the fire and building codes. Was the facility in compliance with the codes? If not, why? Did you previously inform administration that the facility was not in compliance? If you had, copy the memo you submitted to them that informed them the facility was not in compliance. Explain that the failure to comply with the codes may become an issue because the facility was not in compliance with the law.

If, after the fire, the facility is cited for violations, immediately inform administration of these violations. Point out the violations that did not contribute to the fire. If any violations *did* contribute to the fire, the fire department will inform you immediately. Research and evaluate the code violations. Ask the department issuing the violation to explain it to you. Their interpretation of the law may be different from yours. Have your code books accessible when researching their claim. If you feel the violation is unwarranted, put this in your report to your administration along with your reasons.

How Did the Fire Brigade React?

Give administration a synopsis of how your fire brigade reacted during the fire. If they reacted well, give them credit. If there were problems, describe them to administration. Use this report to support or justify budget requests. Stating that the fire brigade did not respond as rapidly as it should have indicates that your department needs more monies allocated for a variety of purposes.

Evaluate Your Plans

Evaluate the fire operation and fire safety plans. In your opinion, were these plans effective during the fire? Were all aspects of the plans followed? If not, why? Should sections of the plans be updated or removed? Ask for input from your staff and other employees who were present when the fire originated. Design a simple questionnaire to determine the following:

Was the fire brigade adequately trained?

Did employees feel safe during the fire?

Did others feel safe during the fire?

Did all systems operate properly during the fire?

Did employees panic during the fire?

Did others panic during the fire?

Was evacuation of the building required?

Was the evacuation handled properly?

How were employees notified of the fire condition?

How were others notified of the fire condition?

Were evacuation procedures followed by occupants?

Did occupants understand evacuation procedures?

Were any occupants trapped in the facility during the fire?

Were stairwells used, by the occupants, for egress?

Was a prearranged meeting place made with the occupants?

These are some questions you may want answered after a fire. Many times, information is overlooked by fire departments and organizations because individuals are not asked specific questions or their opinions regarding a fire.

SUMMARY

It is not easy to be an FSD. You can second guess what went wrong all you like if you were not prepared to handle a fire condition. But this will not change the fact that the fire occurred. If your fire safety plan and fire operation plan were in place, and your fire brigade responded as they were trained, you did one hell of a job. Congratulations!

It is imperative that you learn as much as you can after a fire ravages your facility. If fatalities occur that you could have prevented, you will live with that the rest of your life. If a fatality occurs, even though you had plans and systems in place, you know you did all you could do. FSDs who accept responsibility in fire safety and security will be admired by their peers.

The overall benefits you gain from a post-fire analysis will benefit you and your organization in reviewing and revising your fire operation and fire safety plans. Besides saving lives and protecting property, getting the facility operational in the least amount of time is crucial to the organization. You play a pivotal role in evaluating the aftermath of a fire condition.

After a fire, build a rapport with fire marshals, insurance investigators, and other agencies involved in investigating the cause of the fire, the effectiveness of the fire brigade, and the facility's fire systems. Learn as much as possible from these individuals.

The FSD who conducts a professional and effective post-analysis synopsis of the fire will gain insight into how the staff and facility reacted during the fire. Evaluate all information you obtain. All findings should be presented, in written form, to administration. Analysis results may indicate that a more effective fire system or additional training of personnel is needed. More monies may also be allocated to your department for fire training, hiring, and education.

Glossary

Absolute Radiation: energy radiated in the form of waves or particles which stands apart from a normal or unusual syntactical relationship with other elements

Accelerant: a substance used to accelerate the process for spreading a fire

Aggregate: any or several hard inert materials, such as sand, gravel, or slag, used for mixing with a cementing material to form concrete

AIA: American Insurance Association; originally known as the National Bureau of Fire Underwriters and publisher of the first National Building Code in 1905

Alarm Box: a device that signals and alerts the fire department or alarm company that a potential fire exists in your facility

Alarm Test: a feature incorporated into the design of most fire systems which enables the system to be tested without alerting the local fire department or alarm company

Arson: the malicious or fraudulent burning of property

Atrium: a floor opening or series of floor openings linking two or more stories that is enclosed at the summit of the series of openings and is used for purposes other than an enclosed stairway

Audible Signals: signals capable of being heard during an alarm transmission

Auxiliary System: a supplementary or reserve system capable of functioning when the primary system is nonfunctional

Back Draft: air, smoke, or flames that are moving backward away from the initial fire

BBC: the Basic Building Code; published in 1950; initially prepared and written to discuss construction of buildings and building materials

BLEVE: acronym for boiling liquid-expanding vapor explosion

BOCA: the Building Officials and Code Administration International; publisher of the Basic Fire Prevention Code

Building Codes: codes that are based on model codes or other nationally recognized standards to ensure safer structures

Carbon Dioxide: a heavy colorless gas that does not support combustion and dissolves in water to form carbonic acid

Carbon Monoxide Intoxification: the most toxic of fire gases; it robs the blood of required oxygen and prevents the blood from disposing of the waste carbon dioxide normally brings back in to the lungs

Certificate of Fitness: *C of F;* a piece of paper issued to a qualified individual stipulating that this individual received a satisfactory grade on their on-site examination

Certificate of Occupancy: *C of O;* a piece of paper stipulating that all the criteria subjugated to a specific building, and other germane codes, have been met

CHEMTREC: acronym for Chemical Transportation Emergency Center

Coaxial Cable: a transmission line that contains a tube of electrically conducting material encircling a central conductor that is held in place by insulators and used to transmit telegraph, telephone, and television signals of high frequency

Codifying: to classify or reduce to a code

Combustion: an act or instance of burning

Conduction: transmission through or by means of a conductor

Convection: the transfer of heat by an automatic circulation of a fluid

Convergence Cluster: a group of individuals who, in seeking refuge from a fire, congregate into one specific area of the building they deem safe

Creep: a permanent change in shape due to prolonged stress or exposure to high temperatures

Criticality: highly sensitive policies, procedures, and systems in which the threat of open attack or damage to corporate operations is considerable

DAC: acronym for detection, alarm, and communications

DFSD: acronym for Deputy Fire Safety Director

Emergency Lighting: a backup lighting system that becomes operational when the primary lighting system fails

Egress: a place or means of exiting

Evacuation: the act or process of evacuating

Financial Bond: a written mechanism with sureties, ensuring faithful performance of contemplated acts or duties

Fire Command Station: the nerve center of your facility; where all communications are located and manpower is deployed

Fire Gases: gases that remain when products of combustion are cooled to normal temperatures

Fire Growth: the time it takes for a fire to spread

Fireguard: individual selected by the FSD to protect the facility against fire

Fire Load: the amount of furnishings your facility has in a specific area at the time of the fire

Fire Operation Flow Chart: a chart designed to show others the organization of the fire brigade

FPE: acronym for Fire Protection Engineer

Fire Retardant: material designed or treated to hold back or slow down flames

Fire Stopping: design or materials used to close open areas of a structure to prevent the spread of fire

Fire Warden: an individual in a facility who is given supervisory responsibility regarding a particular floor or area involving a fire or smoke condition

Frequency Division Multiplexing: a signaling procedure identified by the simultaneous transmission of more than one signal in a communication channel; signals from one or multiple station locations distinguished from one another by virtue of each signal being assigned to a separate frequency or amalgamation of frequencies

Gas Tanks: tanks that are fabricated in accordance with Section VIII (Unfired Pressure Vessels) of the Boilder and Pressure Code

Glycerine: chemical used to depress the freezing point of water in portions of a wet standpipe system connected to a public water supply

Gravity Tanks: a tank specifically used for fire protection purposes; designed to store and draw water for a specified facility

Gypsum: a widely distributed mineral consisting of hydrous calcium sulfate; used as a soil amendment and in making plaster of paris

Halogenated Extinguishing Agents: hydrocarbons in which one or more hydrogen atoms have been replaced by halogen atoms; also known as halons

Hazardous Material: materials that could cause temporary disability, injury, or death; moderately combustible and, many times, self-reactive

Heat Conductivity: also defined as the *K-Factor*; the measure of the rate at which absorbed heat flows through a mass of material

Hold Harmless Clause: to protect from loss or liability; indemnify; guarantee

Hydrostatic Testing: testing of fluids at rest or testing of pressures they exert

Hypervigilant: highly alert or watchful of danger; excessive behavior

ICBO: acronym for the International Conference of Building Officials

In-situ: in the natural or original position

Interim: an intervening time; an interval

Integrity: an unimpaired condition

Insulation: a separation from conducting bodies by means of nonconductors so as to prevent transfer of heat or electricity

Ionization: the whole or partial conversion to ions

Jurisdiction: the power, right, or authority to interpret or apply the law

LNG: acronym for Liquified Natural Gas

Loaded Sprinkler: a sprinkler subject to loading or corrosion and in need of being tested or being replaced

Magnesium: a silver-white, light, malleable metallic element that occurs abundantly in nature and is used in metallurgical and chemical processes, photography, signaling, and in the manufacture of pyrotechnics due to the intense white light it produces on burning

Mandates: an authoritative command; a formal order from a superior court or official to an inferior one

Master Keys: complete set of keys to the facility; should be kept in a secure cabinet and issued to the fire department upon their arrival

McCulloh Circuit: the original multiplex circuit used in the security/fire safety business; consists of a series connection of many subscribers via telephone lines

Metallic: of or relating to metal

Nerve Center: fire command center; primary security/fire safety communications center

NBC: acronym for the National Building Code

NFPA: acronym for the National Fire Protection Association

Noncombustible: not capable of combustion

Nonconductivity: a substance that conducts heat, electricity, or sound in very small degrees

Norms: authoritative standards

Noxious Gas: any gas that is physically harmful to living beings

Occupant Load: the total number of persons that may occupy a building or portion thereof at any one time

Ordinary Construction: the type of construction in which outer walls are of noncombustible or limited combustible materials that have a minimum hourly fire resistance and stability rating under fire conditions

Outside Stairs: stairs in which at least one side is open to the outside

PCB: acronym for polychlorinated biphenyl

Photosensitivity: sensitivity to radiant energy

Pilfer: to steal in small amounts and often

RAA: acronym for Risk Analysis Audit

Radiation: energy emitted in the form of waves or particles

Residential Occupancies: structures designed to provide sleeping accommodations

Respondent Superior: a master/servant relationship between two parties

SBCCI: acronym for the Southern Building Code Congress International

Smolder: to burn sluggishly, without flame, and often with excessive smoke

Shrinkage: the amount of merchandise lost through theft or pilferage over a given time period

Smoke Detector: an alarm that automatically activates when it detects the presence of smoke

Smoke Evacuation System: a system that removes smoke, gases, and heat from a building

Sprinkler System: a system designed for guarding a facility against fire or smoke by means of overhead pipes that transport extinguishing liquid to heat-triggered outlets

Statutory: enacted, created, or regulated by statutes

Standpipe: a vertical pipe used to secure a uniform pressure in a water supply system

Standard Fire: a time-temperature correlation

Stipend: a fixed sum of money periodically disbursed for services or to defray expenses

Susceptible: capable of submitting to an action, process, or operation

Suppression: the state of being suppressed

Target Hardening: surveying a physical property and strengthening critical and vulnerable areas

Thermoelectric Sensitivity: relating to or dependent on phenomena that involve association between the temperature of and electric condition in metal or connecting metals

Titanium: a silvery-gray, delicate, concentrated metallic element found merged in ilmenite and rutile and used in alloys (steel) and combined in refractory materials and coatings

UBC: acronym for Uniform Building Code

UL: acronym for Underwriters Laboratories

Underground Structure: a structure or portions of a structure in which the story is below ground level

Ventilation: air circulation

Vertical Opening: an opening through the floor or roof

Visual Signals: signals capable of being seen during an alarm transmission

Vulnerability: weak or inadequate policies, procedures, and systems that seriously affect the safety and security of personnel, assets, and profitability

Water-filled Steel Columns: a liquid-filled, fire resistant column that, during a fire, transfers heat away from the fire location by convection currents within the liquid

Water-surrounded Structure: any structure that is fully surrounded by water

Windowless Building: a building or divisions of a building that do not have exterior portals for ventilation or rescue

Wood Frame Construction: a type of construction in which the exterior walls, wall dividers, floors, roofs, and their foundations are completely or partially manufactured of wood

Bibliography

1. U.S. Department of Commerce and U.S. Department of Health, *Human Behavior and Fire Emergencies: An Annotated Bibliography* (Springfield, VA: Author, 1981), 17.
2. Ibid., 21.
3. Ibid., 23.
4. John P. Keating, "The Myth of Panic," *Fire Journal* (May 1982):57.
5. Ibid., 58.
6. Albert H. Bottoms and Ernest K. Nilsson, "Operations Research," *The Police Chief* (May 1970).
7. National Advisory Committee on Criminal Justice Standards and Goals, *Private Security, Report of the Task Force on Private Security* (Washington: Author, 1976), 197.
8. Ibid., 198.
9. *Fire Protection Handbook*, eds. Arthur E. Cote and Jim L. Linville, Sixteenth Edition (Quincy, Mass.: National Fire Protection Association, 1986), 15–43.
10. Ibid., 15–43.
11. R. Keegan Federal, Jr., and Jennifer L Fogleman, eds., "K-Mart Corporation and Nathan Christian v. Minerva Martinez, Elsa Nichols, and Martha Mata (1988)," *Avoiding Liability in Premises Security* (Atlanta: Strafford Publicatons, Inc., 1989), 144.
12. *The Fire Protection Handbook*, 15–32.
13. *Private Security*, 89.
14. Charles W. Bahme, *Fireman's Law Book* (Boston: NFPA, 1967), 96.
15. *The Fire Protection Handbook*, 15–42.
16. Ibid.
17. Ibid., 16–47.
18. "Interior Fire Alarm and Signal System (Sec 27–968)," *Fire Science Institute Directives* (Sept. 1988):5–18.
19. *The Fire Protection Handbook*, 8–91.
20. Personal interview with Larry Visotsky. 17 Sept. 1990. New York City.
21. Ibid., 16–13, 16–14.

22. *The Fire Protection Handbook,* 16–13, 16–14.
23. Personal interview with Larry Visotsky. 17 Sept. 1990. New York City.
24. *The Fire Protection Handbook,* 17–17.
25. Building Officials and Code Administrators International, *The BOCA Basic National Fire Prevention Code 1984,* Sixth Edition (Danville, Ill.: Interstate Printers & Publishers, 1984), iii.
26. *The Fire Protection Handbook,* 7–174.
27. Ibid., 7–776.
28. Personal interview with William Collins. 25 August 1990. East Setauket, NY.
29. *The Fire Protection Handbook,* 2–24.
30. *The BOCA/National Fire Prevention Code,* iii.
31. Personal interview with William Collins. 25 August 1990. East Setauket, NY.
32. *The Fire Prevention Handbook,* 7–8.
33. Thomas J. Whittle, "The Phases of Partnership," *Security Management,* 34 (April 1990):39.
34. Ibid., 40.
35. *The Fire Prevention Handbook,* 7–5.
36. Ibid., 7–8.
37. R.D. Anchor, H.L. Malhorta, J.A. Purkiss, *Design of Structures Against Fires* (London: Elsevier Applied Science Publishers, 1986), v.
38. Ibid., vi.
39. *The Fire Protection Handbook,* 7–83.
40. Louis Prezetk, *Standard Details for Fire Resistive Building Construction* (New York: McGraw-Hill, 1977), 5.
41. National Fire Protection Association, *NFPA 101, Code for Safety to Life from Fire in Buildings and Structures, 1988 Edition* (Quincy, Mass.: Author, 1987), 101–25.
42. Ibid.
43. Ibid.
44. Jake Pauls, "Development of Knowledge about Means of Egress," *Fire Technology Journal* 20(2) (May 1984):30.
45. "Beverly Hills Supper Club Fire," in an *Investigative Report to the Governor* (September 1977):F–14.
46. L. Bickman, P. Edelman, and M. McDaniel, *A Model of Human Behavior in a Fire Emergency* (Chicago: Loyola University of Chicago, 1977), 32.
47. *The Fire Prevention Handbook,* 7–30.
48. Ibid., 7–49.
49. Ibid., 7–51, 7–52, 7–53.
50. Ibid., 7–54, 7–55.
51. Charles Schnabalk, *Physical Security: Practices & Technology* (Boston: Butterworth-Heinemann, 1983), 88.

52. *Avoiding Liability in Premises Security, a Casebook*, eds. R. Keegan Federal, Jr. and Jennifer L. Fogleman (Atlanta: Strafford Publications, 1989), 37.
53. Ibid.
54. Ibid.
55. Ibid.
56. Ibid., 35.
57. Ibid.
58. Ibid.
59. Ibid., 403.
60. Ibid.
61. *Private Security*, 137.
62. Emmett J. Vaughan and Curtis M. Elliot, *Fundamentals of Risk and Insurance*, Second Edition (New York: John Wiley and Sons, 1978), 413.
63. Ibid.
64. "Eileen Coglianese, as Special Administrator for the Estate of Edmond Coglianese, deceased, v. The Mark Twain Limited Partnership, beneficiary under Trust No. 23904 (1988)," in *Avoiding Liability*, 260.
65. "Arrow Electronics, Inc. v. The Stouffer Corporation, et al. (1982)," in *Avoiding Liability*, 179.
66. *The Fire Protection Handbook*, 3–2.
67. Ibid., 15–41, 15–42.
68. Ibid., 7–11.
69. Ibid., 9–67.
70. Ibid., 8–29.
71. NFPA 101, 101–3.
72. Ibid., 101–166.
73. *The Fire Prevention Handbook*, 3–16.
74. Ibid.
75. Ibid., 8–82.
76. Ibid., 8–88.
77. Ibid.
78. Fire Science Institute, *Hotel-Motel Fire Safety Director's Course, Part 2* (New York, John Jay College: Author, 1986), 37.
79. *The Fire Protection Handbook*, 11–74.
80. Ibid.
81. *The Factory Mutual System Approval Guide* (Norwood: Factory Mutual, 1984), 8.
82. Ibid., 5.
83. Ibid., 7.
84. *Fire Protection Handbook, Fourteenth Edition*, eds. Gordon P. McKinnon and Keith Tower (Boston: National Fire Protection Association, 1976), 13–20.
85. Ibid., 13–24.

86. Willie Hammer, *Occupational Safety Management and Engineering* (Englewood Cliffs, NJ: Prentice-Hall, 1976), 325.
87. Ibid., 321.
88. *NFPA 101*, 101–213.
89. Ibid., 101–8.
90. *Fire Protection Handbook*, 5–3.
91. Ibid., 3–42.
92. Ibid., 3–46.
93. Ibid.
94. Ibid., 3–32.
95. Ibid., 15–2.
96. Ibid., 15–4.
97. Ibid.
98. Ibid., 15–5.
99. Ibid., 1–19.
100. Ibid., 1–21.

Index